THE NEW BREED

THE NEW BREED

. . . . .

WHAT OUR HISTORY WITH
ANIMALS REVEALS ABOUT OUR
FUTURE WITH ROBOTS

KATE DARLING

HENRY HOLT AND COMPANY
NEW YORK

Henry Holt and Company
Publishers since 1866
120 Broadway
New York, NY 10271
www.henryholt.com

Henry Holt ® and ® are registered trademarks of
Macmillan Publishing Group, LLC.

Library of Congress Cataloging-in-Publication Data

Names: Darling, Kate (Research specialist), author.
Title: The new breed : what our history with animals reveals about our future with
    robots / Kate Darling.
Description: First edition. | New York, NY : Henry Holt and Company, 2021.| Includes
    bibliographical references and index.
Identifiers: LCCN 2020052629 (print) | LCCN 2020052630 (ebook) | ISBN 9781250296108
    (hardcover) | ISBN 9781250296115 (ebook)
Subjects: LCSH: Robotics—Human factors. | Human-machine systems. | Human-animal
    relationships.
Classification: LCC TJ211.49 .D37 2021 (print) | LCC TJ211.49 (ebook) |
    DDC 629.8/924019—dc23
LC record available at https://lccn.loc.gov/2020052629
LC ebook record available at https://lccn.loc.gov/2020052630

First Edition 2021

Designed by Kelly S. Too

Printed in the United States of America

1 3 5 7 9 10 8 6 4 2

To my dad, who always encouraged an open mind.

CONTENTS

. . .

INTRODUCTION

"Animals are good to think with."

—Claude Lévi-Strauss

I was twelve weeks pregnant and nauseous, but excited. After two days of co-running a workshop in Mountain View, California, I had been handed an opportunity I couldn't resist, so I woke up at the crack of dawn and flew from San Jose to Denver to Boston to Zurich, and took multiple trains to Bavaria, Germany, determined to get to my destination: Ingolstadt.

Ingolstadt is a university town on the banks of the Danube River with beautiful red roofs and cobbled streets. It's famous for its nineteenth-century medical laboratory, where scientists and students performed experiments on dead pigs, inspiring Mary Shelley to situate a large part of her famous 1818 novel, *Frankenstein,* in this Bavarian city. But Frankenstein wasn't the reason I made the 5,800-mile trek. Ingolstadt also happens to be the home of Audi AG, the German luxury car manufacturer.

Audi had recently launched a research initiative to investigate societal questions around AI, autonomous vehicles, and the future of work, and I jumped at the invitation to attend a meeting in 2017, curious to know what was on their minds. By the time I made it to Audi's base of operations, fueled by adrenaline and excitement, my body was moving into a new stage of pregnancy and my nausea was lifting (thankfully, as the catered buffet lunch in the room was a rich, pungent veal stroganoff on noodles). My visit included a tour of a factory floor where cars were made. It was a gray and cloudy day, and a bus picked us up outside the headquarters where the attendees had gathered and drove us through the drab and massive complex of buildings, dropping us off at a giant warehouse. I tossed my phone into a dirty rubber box in the hallway as instructed and followed our guide onto the factory floor.

In the factory, we marveled at massive cages encasing robotic arms that towered over our heads. The robots swung around and moved through their spaces in a fast, precise, and mesmerizing dance, sparks flying as they worked with the metal pieces that would eventually become cars. As we oohed and aahed over the spectacle, we gave barely any attention to the human workers who were stationed far away in another part of the room, doing something to the car bodies. The smooth operation of the robots seemed routine and almost boring to our guide, which was no surprise. Car companies have been working with caged robotic arms in their factories for decades. But the reason Audi had launched their new AI initiative was because the company knew that these factory robots, despite being an impressive display of high-quality German engineering, were not the robots of the future.

The world of robotics is changing. With increasing developments in sensing, visual processing, and mobility, robots are now able to move beyond their traditional caged existence in factories and warehouses and enter into new spaces—spaces that are currently occupied by humans. Companies like Audi are investing heavily in AI and robotics, not just in their factories but also in their cars. Robots are now being put to work inspecting our sewers, mopping our floors, delivering our burritos, and keeping our elderly relatives company. From our households to our workplaces, a revolution is coming. What does this mean for the people I saw working across the room in the car factory? According to some of the headlines, they aren't the only ones on the cusp of losing their jobs as robotic technology advances: we all are. Against the backdrop of broader economic and social anxiety, the conversation has turned from "Will robots replace me?" to "How soon will robots replace me?"

Many people are not thrilled by the anticipated robot takeover. Our concerns are particularly centered on the idea of creating something like us, with humanlike agency, that will take our steering wheels and harm us or our children. Headlines paint a dystopia of robot brothels and robot-run restaurants and hotels, a world where robots take all human jobs, and where our nannies and boyfriends are replaced by machines. In Mary Shelley's story, Victor Frankenstein studies medicine in Ingolstadt and creates an autonomous, intelligent being that eventually turns against him. Along with the golem from Jewish folklore, Frankenstein's monster is considered an early story about robotics, despite being published more than a century before the word "robot" was coined. Science fiction writer Isaac Asimov would later describe a negative public attitude toward robots as "the Frankenstein complex." Today, a car manufacturer is grappling with a modern version of the narrative that originated in the same city, Ingolstadt, over two hundred years ago.

Is this fear justified? It certainly looks like we're trying to replace people with machines. In the fall of the same year I went to Ingolstadt, October 2017, Saudi Arabia granted a realistic-looking humanoid robot named Sophia Saudi Arabian citizenship. The announcement caused an uproar. A robot was being granted rights in a country that had barely announced (and not yet implemented) women's right to drive cars! I received a flurry of emails and phone calls, especially from reporters who wanted to explore whether robots deserved human rights. At this point, I was very pregnant and ignored most of them. I felt that "citizenship" for Sophia, a robot not nearly as advanced as people imagine, was basically a publicity stunt, but in usual fashion, when robots made the news, I received calls about the legal, social, and ethical issues involved. My own questions, however, centered on why this stunt generated so much attention in the first place.

My passion for robots and society goes back to when I was a law and economics grad student. While pursuing my studies, I met some students from robotics labs, started reading obscure robot ethics papers, and found myself arguing passionately with friends about robots, especially when I'd had a drink or two. I bought a baby dinosaur robot "pet" that I "adopted" (more on this in chapter 10). Thus began my pursuit of questions such as "What impact will increasing robotization have on society?" It was the beginning of a completely different academic career than I had ever imagined for myself. For over a decade now, I've worked side by side with roboticists and applied my legal and social sciences background to the technology. I've researched literature, delved into human psychology, done experiments, and had conversations with people all over the globe.

It's clear to me that the idea of robots we are most familiar with comes from our science fiction. I've always loved science fiction. I grew up reading all the sci-fi I could find, from trashy pulp novels to great authors like Ursula Le Guin and Octavia Butler who opened my mind to new ways of thinking. But now that I work in robotics, I've also seen how our mainstream Western science-fictional portrayal of robots does the opposite. As technology critic Sara Watson points out, our stories, too often, compare robots to humans.

I believe that this human comparison limits us. It stirs confusion about the abilities of machines, stokes an exaggerated fear of losing human work, raises strange questions over how to assign responsibility for harm, and causes moral panic about our emotional attachments. But the main problem I have with our eagerness to compare robots to humans is that it gives rise to a false determinism. When we assume that robots will inevitably automate human jobs and replace friendships, we're not thinking creatively about how we design and use the technology, and we don't see the choices we have in shaping the broader systems around it.

This book offers a different analogy. It's one we're familiar with, and it's one that changes our conversations in surprisingly significant ways. Throughout history, we've used animals for work, weaponry, and companionship. Like robots, animals can sense, make their own decisions, act on the world, and learn. And like robots, animals perceive and engage with the world differently than humans. That's why, for millennia, we've relied on animals to help us do things we couldn't do alone. In using these autonomous, sometimes unpredictable agents, we have not replaced, but rather supplemented, our own relationships and skills.

We've domesticated oxen to plow our fields and learned to ride horseback, extending ourselves and our societies in new ways physically and economically. We've created pigeon delivery systems, set loose flaming pigs to ward off elephant attacks, and trained dolphins to detect underwater mines. From the beginning of laws known to humankind, we've dealt with the question of responsibility when autonomous beasts cause harm, even putting animals themselves on trial for the crimes they committed. And we've also extended ourselves socially: throughout history, we've treated most animals as tools and products, but have also made some of them our friends.

Using animals to think about robots acknowledges our inherent tendency to project life onto this technology, something that has fascinated me for years. From the simple vacuum cleaner roaming around in our physical space, to dragonfly robots that flap their wings in a biologically realistic way, we respond viscerally to moving machines, even though we know that they aren't alive.

In comparing robots to animals, I'm not arguing that they are the same. Animals are alive and can feel, while robots suffer no differently than a kitchen blender. Animals are often more limited than robots—I can train Fido to retrieve a ball, but not to vacuum a floor—but they can also handle unanticipated situations more easily than any machine. The point is that this thought exercise lets us step out of the human comparison we're clinging to and imagine a different kind of agent.

In collecting some of the parallels in the past, present, and future of our relationships to both animals and robots, I've found that using animals to think through our most pressing concerns changes a lot of conversations. Just like animals, robots don't need to be a one-to-one replacement for our jobs or relationships. Instead, robots can enable us to work and love in new ways. Using a different comparison lets us examine how we can leverage different types of intelligences and skills to invent new practices, find new solutions, and explore new types of relationships—rather than re-creating what we already have. Setting aside our moral panic also helps us see some of the actual ethical and political issues we will be facing as we begin to live alongside these machines, from nonlinear economic disruption to emotional coercion.

This book begins with a contemporary exploration of how we are integrating robots into our spaces and systems, drawing parallels to how we've used animals in the past. In this first part, "Work, Weaponry, Responsibility," I pick up many familiar questions that are in the foreground of our conversations about the future: Will robots replace our jobs? Is artificial superintelligence a threat? How do we assign responsibility for unanticipated robot behavior? What I want to illustrate is how much our perception of robots as quasi-humans (falsely) shapes those conversations, and that using an animal analogy leads us down a new path, one that doesn't force us to put productivity over humanity.

The second part of the book, "Companionship," moves slightly further into the future and explores emerging developments in robot companions. Social robots, while not yet widespread, are on the rise. These robots can't feel, but we feel for them, with people even mourning them when they "die." Here, our history with companion animals demystifies the human-replacement stigma around our emotional connections to robots. Recognizing our ability to form relationships with a wide variety of "others" helps us set aside moral panic, but also reveals some unresolved challenges with privacy, bias, and economic incentives that we need to pay closer attention to as we move ahead.

The third and final part of this book, "Violence, Empathy, and Rights," takes the animal analogy all the way into the very futuristic-sounding realm of robot rights. The humanlike machines in our science fiction stories have prompted conversations about our likely future treatment of robots. But looking at the convoluted path of Western animal rights provides a different prediction for how a robot rights movement would play out. Our history of relating to nonhumans shines a harsh and insightful light on how we choose which lives have value, revealing a new understanding of how we relate—not just to nonhumans but also to each other.

Historians and sociologists have long used animals to think about what it means to be human, but animals also have a lot to teach us about our relationship with robots. The robotic technologies that are increasingly woven into the fabric of our daily lives bring questions and choices that we, as societies, will face. This book is a compilation of those questions, those choices, gleaned from the fields of technology, law, psychology, and ethics, and set against a backdrop of our historical relationship with nonhumans, to try and make sense of what a future with this new breed means for us, and how we can shape it.

AUTHOR'S NOTE

WHAT IS A ROBOT, ANYWAY?

"Never ask a roboticist what a robot is."

—Illah Nourbakhsh, a roboticist

Here's a surprisingly tricky question that I get a lot: what is a robot?

We all sort of know what a robot is: the metal Maschinenmensch from the 1920s science fiction classic *Metropolis*; Rosie from *The Jetsons*; and beloved *Star Wars* heroes R2-D2 and C-3PO. My toddler gleefully exclaims "beep boop" upon encountering a robot, like the vintage metal windup toy in his grandfather's office and the robot vacuum cleaner that roams our floor. But he also says "beep boop" to our office printer when it lights up and spits out pieces of paper, and he doesn't think that our computers are robots. The lines that adults draw aren't any less arbitrary. When digital rights expert Camille François and I ran a workshop with a fairly tech-savvy group of colleagues, they struggled to define the term and identify which of their household devices was a robot. Is it a machine that can perform tasks on its own? The dishwasher can do that, and so can a desktop computer, but people hesitated to put them in the robot category on our whiteboard.

Our colleagues weren't being ignorant: the definition of robot is elusive. Coined in 1920 by Karel Čapek, the term "robot" (*robota* = forced labor in Czech) originates from his play titled *R.U.R.* (*Rossum's*

Universal Robots), a story about the exploitation of artificial people who are put to work as "robots" in factories and eventually rise up against their makers. Early on, we started to use "robot" to refer to technologies that replaced humans with machines, applying the term to anything from gyrocompasses to vending machines. Some people say that the definition of a robot is simply a machine that's new and unfamiliar to the general public, and that these robots become "dishwashers" and "automatic thermostats" once the novelty wears off.

Asking roboticists for a concrete definition doesn't help very much, either. Their answers tend to be more technical and narrowly defined, but still leave plenty of fuzzy edges. Most of them agree that a robot needs a body. Artificial intelligence has been a hot topic of discussion for the past few years, but this book is mainly about physical robots for reasons I'll elaborate on in chapter 4—their embodiment has some pretty unique effects.

Some roboticists say that a robot is a constructed system with mental and physical agency that's not "alive" in the biological sense. Others use a paradigm called "sense, think, act," which describes machines that can sense, make autonomous decisions, and act on their physical environments. This sounds pretty good, but it gets tricky when drilling down on what terms like "act" mean exactly. My smartphone has sensors, can make decisions, and act on its environment (by making sounds, displaying light, vibrating, etc.), yet many roboticists don't believe a smartphone is a robot.

Without a concise definition, how can anyone even begin to write a book about robots? I asked one of my most respected friends and mentors, law professor Jamie Boyle, and he responded: "If anyone insists you give them an essential definition of a robot, you tell them, 'Definitions don't work the way you think they do, dumbass'" (the latter word presumably being a term of art in the law). The idea that there could be a definition of anything is a philosophical mistake. Our language is community- and context-specific, something that University of Washington researchers Meg Young and Ryan Calo have demonstrated in the case of "robot": how you define it depends on the field you're in. And that's fine. In fact, the very purpose of this book is to challenge a singular view of robots.

The reason this challenge to our thinking is so important is that robots are unique in a specific way: unlike other new technologies, like a cryptocurrency that people may struggle to picture in their minds, we all have a vivid image of what a robot is. It's an image that's heavily influenced by science fiction and pop culture. This book questions that image, of robots as quasi-humans, and shows that it seeps into how we design and integrate real robots in our world. A lot of the framing here applies to our thinking on artificial intelligence more broadly. At the same time, the ideas here don't apply to every single physical device that could technically be defined as a robot. Instead of establishing perfect definitions and rules that universally apply to all thinking machines, this book encourages us to stretch our minds and question our underlying assumptions.

This exercise begins in Part I in the workplace, where we should be thinking of robots not as our replacements, but more creatively: as a partner in what we're trying to achieve.

THE NEW BREED

I

WORK, WEAPONRY, RESPONSIBILITY

. . . . .

1

WORKERS TRAINED AND ENGINEERED

[Content warning for this chapter: animal cruelty]

"We have used pigeons, not because the pigeon is an intelligent bird, but because it is a practical one and can be made into a machine, from all practical points of view."

—B. F. Skinner

One of the most enjoyable experiences in Claire Spottiswoode's life was the first time she walked through the woods, letting a little bird lead her to honey. Spottiswoode is a zoologist at the University of Cambridge and the University of Cape Town. She's done extensive field research in the savannas of southern Africa, learning how the Yao villagers communicate with a bird called the honeyguide.

The honeyguide is one of the few avians that can digest wax. To access their food of choice, they have evolved to attract the attention of humans, then lead them to beehives. Once people have harvested the sweet, golden honey, the honeyguide gobbles up the exposed comb and grubs. The birds and humans form a perfect team: the honeyguides are far better at finding beehives, which are often located high in the trees, but they need human help to open them.

The honeyguide-human collaboration goes back to at least the 1500s, but some zoologists think that we've searched for beehives together for closer to 1.9 million years. Honeyguides aren't the only animals that

we've partnered with—we've harnessed animals' unique skills to help us with tasks for millennia. Some, like the honeyguide, have evolved in a way that happens to be useful to people, and others have been intentionally domesticated and bred to live and work with us, their entire genetic lineages changed in the process.

The reason we've partnered with animals is not because they do what humans do. We've partnered with them because their skills are different from ours, and because we have much more to gain from combining their strengths with our own. In the same way, technology can and should be a supplement to our own abilities, a way to find the honey we could never reach alone. But this is not how we currently think about robots.

One muggy midsummer day, I stood outside the Baltimore/Washington airport and summoned a ride from the Lyft app on my phone. An older-generation red Prius pulled up nearly immediately and I slid into the back seat, relieved that, despite my delayed flight, I was going to make it to my destination with a few minutes to spare. We cruised down the highway toward Baltimore. My driver, Debbie, was listening to an R&B station, but turned down the volume once the music switched to advertisements so that we could chat. When I told her what I did, she asked the question I've discussed with nearly every driver I've taken a ride with in countless cities and countries over the past ten years: "How long will it take before I'm replaced by a robot?" We spent the next twenty minutes talking about robots and jobs. Debbie, who was close to retirement age, said that she had heard on the news that all human work would be replaced by robots. She was hopeful she could drive for a few more years and retire before it happened, but she was worried for her grandchildren. Suddenly, Debbie realized that the navigation system had failed and we had gone fifteen minutes out of our way. I wound up being late for my panel, but I was glad that Debbie and I had time to chat.

With a big surge of interest in artificial intelligence and robotics in the past few years, the press is eagerly speculating about our future with robots, with headlines like "Will Robots Steal Your Job?," "The Robots Are Coming, Prepare for Trouble," and "Welcome, Robot Overlords. Please Don't Fire Us?" In 2013, a widely promoted University of Oxford study predicted that almost half of all employment in the United States was at high risk of being replaced by robots and AI within ten to twenty years, and others have predicted even greater vulnerability. Technology is advancing at a breathtaking pace, they say. And robots, the story goes, will soon be able to do everything that humans do, while never tiring, never complaining, and working twenty-four hours a day. A 2017 Pew Research study showed that 77 percent of Americans think that during their lifetime, robots and AI will be able to do many of the jobs currently done by humans. According to Pew surveyor Aaron Smith, most people "are not incredibly excited about machines taking over those responsibilities."

Not only are we on the cusp of the robot job takeover, say the headlines; some believe the robots will take over more than our jobs. Artificial intelligence, they claim, is on the threshold of outsmarting us. Respected thinkers have raised concerns about artificial superintelligence, predicting that robots could outpace human intelligence and wreak havoc on the world. From Stephen Hawking to Elon Musk, these high-profile individuals have sounded the alarm on what they view as the greatest threat to humanity, fanning the flames of latent fears. It's easy for people to get on board with the robot takeover narrative, at least in the West. After all, most of our mainstream science-fictional portrayal of robots has been around precisely this topic, from *2001: A Space Odyssey* to *Ex Machina*.

New technologies often inspire concern, but perhaps not quite in the same way as robots. According to tech philosophy and ethics scholars Peter Asaro and Wendell Wallach, our robot narratives throughout history are about good robots turning evil, either turning against their genius creators, like Frankenstein's monster, or turning against human civilization at large. Is this because robots inherently pose this threat? It's worth noting that this fear seems culturally specific. Karel Čapek's famous 1920s play about the uprising of robot factory workers was performed in both Western countries and Japan. But while the West embraced its negative messages in our robot narratives, Japan gravitated toward friendlier robot portrayals in popular culture, like the famous cartoon *Astro Boy*. In the 1960s, Japan began to view robots as a potential driver of productivity and growth, and when robotics played a big role in Japan's economic revival, it inspired a positive image of robots as nonthreatening and helpful to humans.

Many of my colleagues in robotics are weary of the Western trope that the robots will take all the jobs and become our overlords. The news media often reports on their work in ways that are clickbaity and alarmist, complete with an obligatory picture of the Terminator. I've heard curse words directed at the public intellectuals who extol the dangers of robot takeovers, and complaints that the big-name alarmists are mostly physicists, philosophers, and CEOs who don't have in-depth knowledge of artificial intelligence or robotics. But the Cassandras tend to shoot back that the people who actually work in the field aren't the best judges of broader trends. One night at a conference, I watched Sam Harris, a writer and philosopher with a degree in neuroscience, get on a small stage in front of about a hundred roboticists from some of the top research centers in the world and argue that artificial superintelligence was a significant and likely danger to humanity, and that the technologists who disagreed weren't able to see the forest from their position among the trees. The ensuing uproar was monumental.

I was still thinking about his words the next day, as I rested my head against the back of my seat in a fancy black car driving smoothly down the empty early morning highway toward the airport. "What do you think about self-driving cars?" I asked the young, clean-shaven driver in a black suit and tie. He kept his eyes on the road. He told me that he had gone through a year of training to be a professional black-car driver, a lot of which was about more than just driving. He said he was trained to handle unanticipated situations, like protecting his passengers from attacks or violence, and that, if we got into an accident, his first aid skills could save my life. He asked me, solemnly: "Can a robot car do CPR?"

DIRTY, DULL, DANGEROUS

It's not that robots aren't capable or smart—like animals, their physical and sensory abilities are often better than ours. But before I get into the animal world, I want to put the current state of robotics into perspective. Because it's important to understand that robot abilities differ from human abilities in significant ways.

The first practical robot we put to work was a robotic arm called Unimate. Devised by inventor George Devol in the 1950s, Unimate was set up at General Motors in New Jersey to maneuver the blistering hot die-cast car parts that were dangerous for workers to handle. This factory arm was the ancestor of industrial robotics, the technology still used in manufacturing today, and it defined how we would come to view the function of robots in industrial settings.

Robots have classically been delegated jobs that qualify as one of the three Ds: tasks that are dirty, dull, or dangerous for humans. Industrial robots like Unimate ushered in a shift toward automating certain tasks that were high risk or required repetitive grunt work. The machines were accurate, and incredibly strong. Robots could do heavy lifting and take over difficult work in areas with toxic fumes or other health hazards. But they were also fairly crude and limited to very specific tasks, and they themselves were dangerous machinery to be around, necessitating cages and other safety measures to keep humans away.

After the success of welding car parts, the market for industrial robots exploded, as did innovation around what else we could use robots to do. Companies started exploring using industrial robots for tasks like

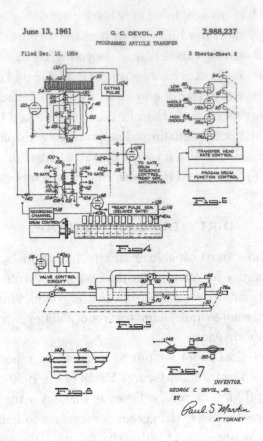

Figures from George Devol's patent on the first robot arm, filed in 1954

packaging, palletizing, basic transport, and loading. Farming industries also got in on the action. On today's farms, agriculture robots spray crops, plant seeds, pull weeds, and even deal with the delicate job of picking fruit. After robots permeated our industrial world, it wasn't long before they made their way into other workplaces.

Straight out of high school, I wanted to work in software development. I interned with a Swiss company, at the time the main IT services arm of a bank, and was delighted with how much the experience matched every pop culture parody of corporate office life. There were lots of lunches and coffee breaks. Nobody seemed to be able to properly explain what the company did, or how their role fit into its larger mission. But my favorite part was the robot. The company had invested in a modern art piece: a robotic office copy machine that was designed to wander the halls, randomly creating and spitting out copies of nothing. I only got to see it once, because, sadly, it wasn't able to recognize stairs and eventually fell down them.

Having a robot in the office was a novelty, but not completely new: beginning in the 1970s, mail delivery robots called mailmobiles were used in office buildings, the first one in Chicago's Sears Tower. The 600-pound, $4 \times 6 \times 2$–foot rectangular robots would move slowly through the halls, ringing a bell as they read barcodes on the floor so that people knew to come collect their mail. (Many FBI offices also used them, as some will recognize from the TV show *The Americans.*) These robots, which weren't phased out until 2016, would beep to alert staffers to their presence, but would often run into people or pin them against the wall. They would get stuck, bump into things, and needed frequent repair.

The technology has improved since the 1970s. Delivery carts shaped like rectangular boxes rumble around hospitals, bringing medicine and other items from room to room. Some hotels have room service robots that can deliver meals, ice, and other necessities to hotel guests, allowing for greater privacy. These robots are able to navigate fairly well-defined spaces and avoid obstacles, stopping or going around people and things instead of bumping into them.

Nowadays, the applications for robots go beyond the three Ds. Robots don't just hold that blistering hot factory part or drill. They are entering our workplaces, households, and public spaces. After the long-unfulfilled promise of widespread robot lawn mowers in the late 1960s, robots are now able to help homeowners cut lawns, as well as vacuum and mop their floors. The machines perform laparoscopic surgery and assist with bone implants, take inventory in stores, and dispense medication in pharmacies. Security robots patrol parking lots, and our military weapons can aim themselves. We're robotizing cars, ships, trucks, planes, trains, and submarines. We already have robots that can drive, mix cocktails, milk cows, and hit ten out of ten free throws. But even though it seems that we're about to be made obsolete, we tend to underestimate our comparative advantages as humans.

Two hospital delivery robots whose sexy nurse names, Roxie and Lola, made me roll my eyes so hard they almost fell out (2012)

HUMANS ARE UNDERRATED

When I first set foot on MIT's campus in 2011, I was eager to see the cutting-edge work being done in robotics. It was fall, and people had returned from summer break ready to do research and present their work. But while everyone was happy to tell me what experiments they were running and explain all the technology they were using, my requests to see demos were mostly denied. "We can only turn this robot on for testing," they would say, or "these are all broken right now, but we can show you a video." Some of the more well-known robots that I had been excited to finally see in person had been out of order for so long that the only graduate students who knew how to repair them had long left the Institute, taking their knowledge with them. While industrial factory arms and newer commercial robots are more robust, this sobering vision of robotics at MIT isn't uncommon.

It's no wonder my colleagues, who work for months to get a simple demo to function, roll their eyes at the worry that robots will replace us. The reality is that we are in the process of creating a huge range of different types of robots. These robots, while a far cry from our science-fictional depictions of machines, have a lot of strengths. But they also have limitations.

In January 2019, a dam at a mine in Brumadinho, Brazil, collapsed, releasing a mudflow that poured through the facility and surrounding area, killing 270 people. Mining, still a necessity in most countries, is one of the most dangerous jobs in the world, according to the International Labour Organization, but this is gradually changing as more companies recruit robots to help. From executing drilling plans to detecting gas leaks to removing loose rock that could pose a hazard, autonomous and semiautonomous technology is able to take on some of the risky business of mining. In Pilbara, a sparsely populated area in Western Australia, driverless robot trucks carry iron ore across the deep red sand plains. The trucks belong to Rio Tinto, the second-largest mining company in the world (probably better known for their destruction of an ancient indigenous site in 2020). The company signed a deal to expand its fleet and have 130 of the enormous, autonomous transporters at work by 2021.

Though the machines in Pilbara look like they're operating all on their own, Rio Tinto actually just shifted the human work to Perth, Australia, nearly one thousand miles south of the mines, where a team of people coordinates and monitors the robots from an air-conditioned control center. Shaniel Davrajh, a principal engineer at the Council for Scientific and Industrial Research (CSIR) in South Africa, admits there is no "silver bullet" to replace people in mining. The most promising path is to incrementally create tools that help the miners themselves work more safely and efficiently. While these developments may change the staffing needs in mining companies, they also dramatically improve working conditions in a dangerous and historically exploitative industry.

Even in the world of dirty, dull, and dangerous, where the ideal seems like it would be to replace the human, what often happens is that robots move the human to a cleaner or safer position. For example, robots have been used for explosive ordnance disposal for decades, defusing bombs and detecting land mines. Working with a partly autonomous tool lets people stay out of harm's way while evaluating the situation and context (more on these robots in chapter 6).

The mining truck operation is similar to the US military's Predator drones, which are flown by people sitting thousands of miles away. Semiautonomous piloting isn't very new; we've long automated large parts of flying commercial aircraft. But despite having automated technology in our planes for decades, we still put pilots in the cockpit. Even though today's robots have autonomous capabilities and can be sent on missions by themselves, they almost always have a human operator in the loop. This helps compensate for the fact that our robots are far from being able to do it all on their own, but it's also often a better arrangement than replacing humans.

People are more talented than we give them credit for. Despite our imminent future of self-driving cars, my Baltimore driver Debbie still has a lot of time if she wants to retire late. Even for the mapped-out and highly traffic-rule-oriented environment of a street, we overestimate how soon human drivers can be made completely obsolete. Self-driving cars are being tested, but are still elusive, as programmers struggle to account for the extremely long tail of rare and unexpected occurrences on roads, from chipmunks to plastic bags. There are many things that can happen, even on a quiet street, all different things and all unlikely, but so many of them that they're nearly impossible to avoid in aggregate. Drivers in Boston, Massachusetts, are called Massholes for their unpredictable, erratic, and extremely aggressive behavior, and navigating Boston driving is a cakewalk compared to the traffic of Mumbai.

Today, little robots roll around the sidewalks of the Bay Area in California to deliver food. But unlike a hotel delivery robot that has a very well-defined environment, these robots aren't able to manage the city streets on their own, so they are remote-controlled by humans. Another Bay Area–based robotics company recently announced preorders for a robot "that can do a variety of household chores along with washing, drying, and putting away the dishes." The headlines don't reveal that the robot is driven remotely by human operators, who perform the tasks via cameras that let them see what they're doing. The company claims that the robots will learn to do more by themselves over time, making the human operators "less needed," but that's quite a stretch.

Robots aren't good at handling navigation in complex areas with a high number of unanticipated occurrences. In spaces where things are more predictable, say, warehouses where the robots can follow markers on the floor, or a trucking highway in the middle of a desert, autonomous vehicles show great promise. Still, a complete handoff is rare for a reason: robots don't always do so well when left completely to their own devices.

Elon Musk, CEO of electric vehicle company Tesla, had long argued that we should embrace technology in order to do away with human workers in industry. But when he decided to create a completely autonomous assembly line in his Silicon Valley factory, he ended up in what he described as "manufacturing hell." Musk had promised to produce five thousand Model 3 electric cars per week in 2018, but Tesla couldn't even make half of them. What went wrong?

According to analysts, the robots, while able to work consistently and precisely, weren't able to recognize the litany of minor defects that can happen during the manufacturing process—slightly crooked parts, for example—leading to problems down the line. Human workers have the flexibility needed to recognize and correct for unexpected errors in the assembly process, which is particularly crucial during the final assembly of a car. In fact, other car manufacturers like GM, Fiat, and Volkswagen, had all previously tried—and failed—to automate final assembly. The analysts concluded that "automation simply can't deal with the complexity, inconsistencies, variation and 'things gone wrong' that humans can." Musk had to acknowledge that his goal of complete automation wasn't achievable. On April 13, 2018, he tweeted that human workers were underrated.

Today, there are robots in research labs all across the world that come in all shapes and sizes. Their capabilities are varied, and their uses are broad. The technology already permeates industries like factories, farming, mining, and sea and space exploration, and robots are increasingly appearing in our homes and healthcare facilities. But the skills

Elon Musk ✔ @elonmusk · Apr 13, 2018   ⌄
Yes, excessive automation at Tesla was a mistake. To be precise, my
mistake. Humans are underrated.

  💬 1.2K   ⟲ 8.5K   ♡ 39K   ↑

and abilities of these robots are vastly different than our science fiction visions, and that's largely because artificial intelligence is not like human intelligence.

While robotics is progressing, it's noteworthy that we've been developing and using factory robots for half a century yet haven't managed to fully automate the production process. This is because robots are very good at doing focused, well-defined tasks but don't understand context and can't handle new situations. A robot built for welding can't task switch to picking up a loose part that falls on the floor elsewhere, something its human coworkers can do effortlessly. Our advancements in artificial intelligence, as amazing as they are, haven't gotten anywhere near understanding how to create the adaptable, flexible general intelligence that a human, even a toddler, has.

Many years ago, a friend invited me to a costume fundraiser. When I got there, she told me that she had also invited her neighbor, an individual named Rodney Brooks. I stood near the door in my dinosaur costume all evening, eager to meet him. Finally, a sixty-year-old man arrived at the door, and I recognized his swoops of wavy hair. I quickly blurted out my planned line, "Wow, you're dressed like Rod Brooks!" Confused, he politely asked me whether we knew each other. "No," I said, "but you're famous," and then I just stared at him. He told me he lived next door. "I know!" I said excitedly. Our hostess interrupted us, and I fled, spending the rest of the night in post-social self-recrimination. The following year, I saw Brooks at a conference. "Please don't remember me, please don't remember me" pleaded the voice in my head, as I saw him walking toward me. "Aren't you the lady who was wearing a dinosaur costume?" he asked.

Why was I so starstruck to meet my friend's neighbor? Because if anyone knows robots, it's Rodney Allen Brooks. He revolutionized the field in the late 1980s by demonstrating that robot intelligence didn't need to be made of abstract mathematical models. Instead, he argued that robots could explore the world using their senses, like animals, and process the data they collected. Brooks built countless robots, directed the Artificial Intelligence Lab at MIT for a decade, and cofounded the robotics company iRobot. (He's also championed women in robotics, with many of his former students becoming robotics legends themselves, including iRobot cofounder Helen Greiner, Google X cofounder Yoky Matsuoka, and MIT professor Cynthia Breazeal.)

Brooks isn't what anyone could call a technophobe—he believes that robotic technology is incredibly useful and has dedicated much of his life and career to proving that. But again and again he's criticized the persistent belief that, because a robot is good at doing one thing, it is smart enough to do other things. And he's experienced firsthand the challenges of getting robots operational enough to perform one task well. As Musk learned the hard way, we tend to underestimate the difficulty of automating most processes. A robot, says Brooks, is incredibly narrow in what it can do. The idea that this will somehow dramatically change anytime soon is faith-based, not science.

Brooks believes that robots aren't anywhere close to replacing us, but can be effective collaborators, so he founded a company to build robots that work together with humans. Rethink Robotics didn't survive, but the field of collaborative robots did. Today, companies and research groups are developing robots that work in industrial settings like factories but are safe enough to interact with human workers, because that interaction is the real future.

It's clear we haven't been able to reach the goal of replacing people, but it's also questionable whether that's the ideal goal. In most cases, the outcomes are better when robots work with people. In well-defined spaces, like rows of crops, robots are able to take on more of the work. In other areas, like delivering a hamburger in San Francisco, they need a ton of human help to deal with the unexpected. Rather than viewing these limitations as a tricky phase on the way to human replacement, we should stop and ask why we are trying to re-create human skills at all. Why are we trying to replicate something we already have?

The more fruitful path is to explore what else we can come up with. Where robots truly shine isn't in replacing the college student who delivers pizza. They're most powerful when their form and function enable them to help us with things we can't do very well ourselves, or even at all. It's here that animals provide a better framework for examining our relationship with robots. That a human-robot tandem is the actual ideal becomes clear when we look at how robots, like animals, can supplement human abilities.

Throughout history, we've drawn on animal skills for many purposes. We've used animals in transportation, hauling, discovery, espionage, communications, and even as weaponry. As we've recruited animals for an increasingly broad array of jobs, we've drawn on a wide range of strength, speed, physical forms, and senses that each supplement our own abilities. Along the same lines, robots and their unique abilities can open even more doors, the most important of which will require us to join forces and combine our respective talents.

EARLY PARTNERSHIPS

The first domesticated animal in history was the dog. Descended from wolves, dogs were the first creatures we partnered with to achieve our goals. Early on, they helped humans hunt, pulled sleds, were used as a source of food, and eventually became our companions. Similarly, cats were initially domesticated as hunters, helping us to catch the rodents that were getting into our food sources. The Roman army lugged cats with them on their travels to guard their grain. Today, a lot of our cats play with $7 rubber mouse toys instead of working to keep our homes vermin-free. I'll talk about dogs and cats in more detail in chapter 5. Although dogs were the first, they arguably weren't the most impactful type of animal domestication. While it was helpful to find hunting partners in dogs and cats, the real shifts in the meaning and transformative power of animals happened once animals enabled farming and transportation, changing our landscapes and societal structures in fundamental ways.

Arguably the most important domesticated animal in the world is the ox, a plow-pulling, load-bearing grunt worker of a beast. Descended from aurochs—wild oxen, six feet tall at the shoulder, whose tanklike bodies and massive horns appear in cave art throughout Europe, Asia, and North Africa dating back 2 million years—modern-day domesticated oxen changed the game for Mesopotamians. Appearing around the fourth or fifth millennium BCE, domesticated oxen replaced the hoes, sticks, sweat, and muscle that earlier farmers used to plow the earth. (The origin of the measurement unit "acre" is the amount of land an ox can plow in a day.)

This refrain plays on repeat in societies throughout history and across the world as animals ranging from llamas to reindeer have found themselves hauling our gear and transporting us from one place to another. On our honeymoon, my husband, Greg, and I drove through the middle of Australia, from south to north, following the route of the original Ghan railway. The empty expanse of the Australian desert is impressive. We had to pack gallons of water and canisters of gas in our off-road truck to make sure we could make it to the next roadhouse, the small stations where truckers fill their tanks and eat a hamburger or a toastie (a type of grilled cheese, in this case with meat fillings because vegetables were nearly nonexistent). Given the massive distances and endless empty space along our route, people would never have been able to traverse the country, let alone build a railway in 1878, without some serious nonhuman help.

The Ghan railway, originally called the Afghan Express, was named after the Afghan camel drivers who first crossed through the center of Australia in the nineteenth century. Contrary to what I learned as a child, camels do not store water in their humps, at least not directly. Their humps store fatty tissue, which produces water when metabolized. This

Plowing with a yoke of horned cattle in ancient Egypt, ca. 1200 BCE

lets them go as many as ten days without water, even in the heat of a desert. Camels are also sturdy enough to survive fluctuations in temperature that would kill other animals. Their wide "toes" (part of their hooves) are ideal for traveling on sand. Once we domesticated this magnificently resilient humped creature, we could connect entire continents together, creating inroads from Africa to Asia to Europe.

We've also benefited from having a wide variety of different animal superpowers to choose from. Donkeys have a bad reputation for being stubborn, but they are also resilient, so they've accompanied eastern Mediterranean traders and travelers, hauled loads for construction in Greece, and were industrially bred by the Romans as work and draft animals. And then of course, there were horses, which were much faster than some of our other draft animals, allowing us to go farther where we had flat surfaces. Long before we had steam-powered vehicles, let alone semiautonomous cars, we hitched horses to carriages and moved around our world in completely different ways.

Like the remote mining trucks that carry ore in Australia, human-animal teams have long allowed for more efficient forms of transport. In fact, before we had robots to help us in mining, we used pit ponies

Lionel Walter Rothschild drives a zebra carriage in London (1895). Zebras are aggressive and very difficult to tame.

(actual ponies but also other small and sturdy pack animals in mines) as the original autonomous ore haulers. Pit ponies pulled carts of materials through and around mines up until as late as 1999.

Beyond these early partnerships in farming and transport, there are numerous other ways that we've used animals to do things for us in our skies, seas, and gutters, from ancient times to the present day. And, as much of the following collection of examples shows, many of these uses have parallels to the current and future applications for robots. (If you're already convinced and have no interest in this smorgasbord of animal and robot stories, feel free to skip ahead to chapter 2 but note that you will miss out on the killer dolphins.)

THE ORIGINAL AUTONOMOUS WEAPONRY

Like robots, we've often drawn on animals for jobs that are dangerous, tasking them with everything from medical testing to space exploration. In 1957, the Soviet Union put Laika the dog on their Sputnik 2 rocket, to see what would happen. Laika was preceded by fruit flies, which the Americans shot into space in 1947, and a rhesus monkey named Albert, who blasted off in 1948. Only the fruit flies survived their adventure. Since then, numerous animals have gone to space. But speaking of dangerous jobs, we've shot animals not just into orbit—they have also served as our weapons, sometimes in very similar ways to how we use robotic technology in war.

In his book *Wired for War*, political scientist and twenty-first-century warfare specialist P. W. Singer details our technological progression from land mines, remote-controlled tractors, and boats carrying thousands of pounds of explosives, to smart bombs like cruise missiles that can guide themselves with target recognition software. We've used guns that can find and take out incoming rockets or artillery, and we're developing remote-controlled ground robots that can carry a machine gun and four grenade launchers, as well as help guide their human operators in aiming at targets. Today, some experts are concerned about the use of autonomous or semiautonomous machines in warfare, and for good reason. But long before we had these semiautonomous weapons made of wires and metal, we used autonomous weapons made of fur, feather, and bone.

When the Greek Megarians were attacked by Indian war elephants around 270 BCE, they set flaming pigs loose to terrify the giant yet skittish gray beasts. According to historian Adrienne Mayor, using animals as "biological weapons" in war was risky because their behavior was both unpredictable and uncontrollable. But that didn't stop ancient armies from using everything from stampeding cattle and elephants to bee and scorpion bombs.

In ancient times, dogs were trained to attack both people and horses in combat. The Romans even gave their war dogs armor and spikes. In World War I, a few countries had canine corps, but the American military's only use of dogs was some huskies in Alaska. This changed in World War II, when the United States started making more extensive use of dog soldiers. The American dogs started out guarding domestic facilities, but when a civilian organization called Dogs for Defense began to call for people to donate their dogs to the war effort, effectively drumming up 19,000 puppy recruits, the military began to train them for a wider range of hunting, guarding, and sniffing services and officially started the War Dog Program.

Dogs have even been used as living bombs, to mixed effect. Declassified documents show that the US military strapped explosives to dogs in World War II, but the most infamous use of dogs as bombs is the failed Soviet anti-tank dog attempt in the 1930s. The dogs were initially taught to place bombs underneath tanks and then return to their handlers, but when this turned out to be too complicated, the Soviets trained them for suicide missions. Unfortunately, because they made the mistake of having the dogs practice on their own tanks, it meant that, come crunch time, some of the dogs dutifully took out Soviet tanks instead of German targets. Unable to practice under real battle conditions, some dogs were startled by the sudden sound of gunshots and raced back to their handlers, live explosives and all.

The anti-tank dogs illustrate that autonomous weapons aren't new, and also that caution is warranted, because it can be hard to anticipate every possible outcome. Animals and robots can fly, swim, and sneak around on our behalf, but their unpredictable behavior means that using them isn't always easy. The CIA tried to get cats to serve as Kremlin and Soviet embassy spies in the 1960s, going so far as to implant microphones

United States Coast Guard dog, WWII

and radio transmitters into their bodies, but apparently the felines were too distractible (or didn't care enough) to successfully perform secret agent missions. This feline failure hasn't stopped us from continuously experimenting with autonomous animal weapons, the attempt to create a "bat bomb" being another infamous example.

During World War II, before automatic guided missiles were invented, the US Army started testing bomb casings that could hold a thousand bats. The idea was to strap incendiary devices onto each individual bat, and then release the "bat bombs" over cities in Japan, where they would flutter around and settle on wooden roofs and buildings, burning the infrastructure to the ground with thousands of fires. In 1943, after spending over $2 million and testing six thousand bats, the army had only managed to get the bats to set a few things on fire, their own airfield hangar being one. The program was soon abandoned.

Another abandoned World War II autonomous weapon system was "Project Pigeon." Notorious psychologist B. F. Skinner convinced the American National Defense Research Committee to give him $25,000 to test a pigeon-driven missile system. His plan was fourfold. Step one: Train pigeons to recognize and peck at an image of an aerial target. Step two: Strap them facedown into a bomb built with special outward facing lenses that could project an image of the outside world near the pigeons' beaks. Step 3: Attach directional sensors to the bottom of each plate, and connect them to the bomb's fins. Step 4: The bomb drops, the pigeons peck at the plate when the targeted image comes into view, and the fins steadily guide the bomb in the direction of the pecks. (To reduce error, each bomb had space for three trained birds.) Skinner debuted the system to mixed reviews from military researchers, and the program was canceled. Shortly afterward, a research group developed the "Bat" bomb, this time using animals as an inspiration only. They equipped a missile with echolocation abilities, and it became the first autonomously guided bomb used in war. While our technologies differ from animals in many ways, the parallels in how we use them as extensions of our abilities (for better or worse) are stark.

THE ORIGINAL DRONES

Drones, more accurately known as UAVs (uncrewed aerial vehicles), were originally developed for surveillance and discovery, able to capture broad aerial imagery on behalf of those on the ground. They are used in war to shoot missiles, but recreational, commercial, and research UAVs now vastly outnumber military UAVs, helping us with search and discovery, data collection, infrastructure inspection, and delivery.

Having eyes in the sky is useful to us, as is the ability to deliver objects by air. When lifeguards in Lennox Head, Australia, were in a training session to learn how to operate a new drone called the Little Ripper, they received a call that two swimmers were struggling in powerful, nearly ten-foot waves far from shore. The lifeguard who was test piloting the drone took action: he located them with the Little Ripper, and the drone dropped its rescue flotation device into the water. The swimmers were saved, and faster than the lifeguards themselves would have been able to rescue them.

Amazon has announced large investments in a future of drone product delivery, with the promise of getting packages to consumers within thirty minutes. This announcement has been mocked by some of my colleagues, who imagine the drones trying to navigate landing in cities, or getting tangled in power lines, shot with rifles, chased by dogs, and injuring people when they malfunction and fall from the sky. But UAV delivery is not an impossible proposition. While it may not be able to happen in New York City yet, drones are already starting to become a useful resource in remote and difficult-to-access areas that need critical health supplies.

Our ability to see and deliver things by air also has a long history. In 1907, a German pharmacist named Julius Neubronner was using pigeons to deliver medicine, just like we're beginning to do with drones today. In experimenting with various forms of pigeon transport, Neubronner also invented a pigeon-operated camera that could take photos while the bird was in flight. Before we ever dreamed of being able to order a quadcopter UAV on the internet for our aerial photography hobbies, the CIA started putting spy cameras on pigeons. And long before that, pigeons were in widespread use for another purpose: transporting messages.

The Persians and Egyptians used pigeons to carry messages as early as the sixth century BCE, and the pigeon has held this role in communications from then until recent history. Before we had radios, carrier pigeons delivered reports on the news, sports races, and other events, and they were even formally used as mail services.

When Paris was under siege in 1870–71, with all communication lines cut, it enacted its fallback plan: using a pigeon post system to bring in messages from outside the city. (The problem with pigeons is that they only work in one direction, so the Parisians had to fly the pigeons back out of the city by hot air balloon.) The Prussians responded by using hawks to try to catch the pigeons, but because messages were sent in multiple copies, they couldn't halt all pigeon communication.

The pigeon isn't the only bird that helped with siege deliveries. During the Spanish Civil War in the 1930s, Captain Cortés's guards were under siege in the monastery Santa María de la Cabeza and in need of supplies. Nationalist pilots dropped live turkeys out of aircraft as "parachutes." The turkeys delivered the supplies into the monastery by flapping their wings to slow their heavily loaded descent until they reached the ground (where they themselves became part of the food delivery).

JULIUS NEUBRONNER, 1852–1932

The original aerial photography UAVs (1909)

Following the Prussians' example in the Paris siege, some countries attempted to train falcons to take out wartime messenger pigeons, and the Chinese responded by outfitting their pigeons with bells to scare away birds of prey. Today, using birds is still a tempting option, as the French air force and Dutch police forces are experimenting with training eagles and other birds of prey to hunt down and neutralize small drones.

Delivery by air, rather than a new invention, has been a vital part of our history. Hundreds of thousands of pigeons delivered messages during the First World War. In World War II, British forces came up with a plan called Operation Columba, for which they started to airdrop boxes of homing pigeons into German-occupied territory, with the idea that members of the civilian resistance could send the pigeons back with intelligence information. The Germans responded by airdropping their own pigeons and pretending that they came from the Brits, asking for the names of local allies. Shortly after this pigeon fiasco, the birds performed a crucial function during the D-Day invasion of Normandy, when they delivered the Allied forces' messages about the progression of the operation.

THE ORIGINAL BODY EXTENSIONS

One of the huge benefits of animals and robots is that they come in many different sizes, shapes, and forms, allowing us to literally extend our own physical embodiment. Robots, like animals, are useful in getting to places we can't get to on our own, like the skies, the surface of Mars, nuclear or earthquake disaster zones, or seas too rough for humans. But using animals as extensions of ourselves has been useful not just in places we couldn't travel to on our own, but also places that our body shapes can't access—narrow spaces, like sewers, gutters, and pipes. The ferret is an adorable example of using animals to do things we can't. In ancient times, and in a few places to this day, the slender, curious animal was put down rabbit holes to scare out the bunnies.

In the United Kingdom, ferrets—a group of which is literally called a "business"—help humans locate breaks in underground pipes and thread electrical cables through intact pipes meters below the Earth's surface. In the 1970s, physicists at the National Accelerator Lab (now Fermilab) in Batavia, Illinois, sent a diaper-clad ferret named Felicia scurrying around their $250 million particle accelerator to clean the instrument's narrow and near-infinite vacuum tubes.

Felicia was eventually replaced by a machine—a "magnetic ferret"

To: Chris Couch · Details

oh yeah, there's a whole weird history of ferrets doing crazy jobs

These run cables in buildings that are historically protected. In those buildings, you can't like rip up the floorboards to run fiber optic cable or anything, so they send ferrets with tiny GPS backpacks and a drawstring attached to their harnesses through the pipes

Omg

There was a ferret used to clean the National Accelerator Lab's particle accelerator

and one used to run cable at Princess Di's wedding

iMessage

designed by engineer Hans Kautzky—as have most other working ferrets. We have new robotic technology that allows us to create modular and even soft bodies that can function like ferrets or snakes—or something else entirely. Some robots are able to assist us with inspecting, cleaning, and even repairing ducts, pipes, and sewers, which is particularly useful in tight spaces and when a system can't easily be dismantled for human inspection.

In healthcare, robots help directly extend the physical capabilities of surgeons, guiding them to do a more precise job and allowing for less invasive procedures. Today, telepresence robots make remote specialist diagnoses possible by letting doctors be in two places at once. Like the guide dogs (or miniature horses) that assist humans who have difficulty seeing, therapy and rehabilitation robots are assisting with exercises or other physical tasks like walking and carrying, while body enhancements like robotic suits or limbs help people with motor activities.

Despite the persistent belief that typical robots look humanoid, we are actually creating robots that swim, fly in the air, scale walls, and crawl, slither, or drive on the ground, providing us with tools that we can use for wonder or warfare all over this enormous planet. And the usefulness of robots doesn't end at their various physical forms. Like animals, robots are also able to sense the world around them in ways that are completely different, and supplemental, to how we do it.

THE ORIGINAL SENSORY EQUIPMENT

Many animals have superpowered noses. Truffle pigs locate mushroom delicacies for us, navies have employed cats and even ferrets as rodent control on ships, and dogs alert us to drugs, bombs, and even electronics in airport traveler luggage. Because dogs can detect and distinguish between faint scents, even at a distance, we've trained them to help us find things we can't find on our own, including people. St. Bernards, which were originally domesticated as farm dogs, were trained by monks in the Swiss Alps to search for lost travelers, especially those caught in avalanches.

Dogs have guarded us for thousands of years, alerting us to unwelcome intruders—and sometimes welcome ones, like the mail carrier—and they join a long list of creature sentinels that includes geese and some sea mammals. Predatory species like dogs, cats, and falcons have greatly expanded our abilities to locate, catch, and retrieve prey. In northwestern Mongolia, some of the nomadic Kazakhs have for hundreds of years cultivated a practice of capturing and training golden eagles to help them hunt red foxes and wolves in the Altai Mountains.

Our history and current practices of using animals for search and discovery parallels some of our ideal use cases for robots. For example, robots have had some recent success in helping us look for bombs, and we've also long trained dogs for this type of military use, from finding trip wires and mines in North Africa, to exploring tunnels in Vietnam, to detecting explosives in Iraq and Afghanistan. In some areas, giant mine detection rats (MDRs) have proven even more effective than dogs at helping us locate and clear land mines. They are easier to train and handle than dogs and too light to trigger the mines themselves. The rats are released onto land mine fields in Mozambique, where they carefully sniff around for explosives and alert their human handlers. Unlike metal detectors, which go off for every random piece of buried scrap metal, the rats are trained to detect certain smells and can work faster than any human with standard tools.

We've also used bees, elephants, and even specially engineered

Eagle hunter

grasshoppers for explosive detection, as well as using the unique sen-
sitivities of animals to keep us safe in other ways, similar to how we
use machines. A Czech beer brewery in South Bohemia named Protivin
pumps their brewery water source into tanks with crayfish in order to
ensure water quality. By monitoring the animals' movement and heart-
beats, they can detect changes in the water purity prior to brewing their
beer.

Of course, the most iconic example of using sentinel species to
warn us of invisible hazards is the canary in the coal mine. Canaries
are more sensitive to carbon monoxide than humans and most other
animals because they breathe in so much oxygen to power their tiny,
flight-capable bodies. In the early 1900s, the British physiologist John
Scott Haldane suggested that the little birds could be useful in mines,
and his idea quickly caught on as practice. For decades, miners brought
canaries with them to test the presence of dangerous fumes, until the
feathered detectors were gradually replaced by electronic sensors in
the 1980s.

Animals still often outperform robots, but they do have some draw-
backs. During the Gulf Wars in 1990 and the early 2000s, the US
Marines came up with Operation Kuwaiti Field Chicken (KFC)—a
plan to bring in truckloads of poultry with the idea that the fowl
would serve as egg-laying early-detection systems for incoming chemical gases. Unfortunately, Operation KFC never really took flight, so
to speak, since most of the chickens immediately died of unknown
causes long before they had the opportunity to die of possible gas
poisoning.

Today, we are creating robots that provide us with tools that we can
use for discovery in the air and on earth, and we are developing modern sensors, biomechatronics, and wearable technology to enhance
our physical abilities, just like we've long used the sensory capabilities
of animals to extend our ability to perceive the world around us. This
litany of animal examples shows how we could and should lean into
the supplemental abilities of robots, rather than focusing on the idea
of automating human tasks. A final example that directly illustrates the
parallels of these nonhuman partners is how we use robots and animals
underwater.

A canary, used to detect carbon monoxide gas in mines (1928)

THE ORIGINAL UNDERWATER AGENTS

The Thwaites Glacier, also known as the Doomsday Glacier, in Antarctica is one of the most inaccessible places in the world. It's also one of the world's fastest-melting glaciers. At 74,000 square miles, it's about the size of Florida or Great Britain, and it performs the crucial task of slowing the ice behind it from entering the ocean. The Doomsday Glacier is on the verge of collapse, which will cause the global sea level to rise by almost three feet, creating flooding, shoreline erosion, and loss of islands, and threatening coastal life everywhere. When do we expect this to happen? In order to make predictions, scientists need to be able to look at the structure more closely, including the grounding zone: an area hundreds of meters below sea level under the Doomsday Glacier, where the ice shelf is supported by rocky land.

In 2020, an autonomous underwater vehicle (AUV) named Icefin made this task possible. Designed to propel themselves forward in water, AUVs carry cameras and sensors, like sonar, magnetometers, and compasses, all of which let them navigate and map their environments. Researchers from the US and the UK fought through brutal weather conditions for two months in order to feed the yellow, torpedo-shaped robot down a nearly half-mile-deep borehole in the ice. From there, Icefin swam on its own for over a mile until it reached the grounding line. Icefin mapped and measured melting conditions at the base of the glacier, giving researchers an unprecedented glimpse at the grave effects of climate change in one of the most structurally vital points on Earth.

Icefin isn't the only robot helping gather information about Antarctica's glaciers. Researchers have also used AUVs to find critical warm water pathways that could accelerate melting. With climate change–related dangers looming on the horizon, humanity needs all the help it can get, and bolstering research can be a matter of life and death.

The global market for autonomous underwater vehicles is expected to grow to $550 million by 2025, as AUVs are increasingly used to help with scientific research, seabed mapping, detecting hazards, exploring ship and airplane wreckage, retrieving lost equipment (like the missing airplane black box from crashed Malaysia Airlines Flight 370), and a variety of other government, research, and industry endeavors. And just as AUVs are deployed for planet-saving purposes, they're also used for the opposite. To the frustration of conservationists, who use AUVs to study everything from bleaching coral reefs to sea sponge ecosystems miles beneath the surface, AUVs are being put to work mining the rare earth metals in the ocean that are used in our iPhones.

As an MIT colleague of mine likes to say, there's a fine line between search and rescue, and search and destroy. The Russian firm Region has developed a military AUV that can be armed with a special pistol—for detonating mines, they claim. AUVs have a history of military use going back to the 1970s, when both the US and Soviet navies were racing to develop underwater drones. It took a while for the technology to be useful, but eventually AUVs were used to help find and clear mines buried on the ocean floor. In our praise of—and worry about—these emerging technologies, it's worth pointing out that ocean mine detection isn't a new job at all, nor is the Russian idea to mount a weapon on an autonomous underwater entity.

In the early 2000s, a Russian man named Boris Zhurid struck a deal to sell the Iranians a large collection of weaponry. He chartered a transport aircraft to make the delivery from Sevastopol, the largest city on the

Crimean Peninsula, in the Black Sea, to the Persian Gulf. An old sonar manufacturer brochure describes what Zhurid was peddling as "self-propelled marine vehicle[s], or platform[s]; with a built-in sonar sensor system suitable for detecting and classifying targets; and carrying an on-board computer . . . capable of being programmed for complex performance." The cargo of Zhurid's chartered plane? Twenty-seven animals, including dolphins, walruses, sea lions, seals, and a white beluga whale.

Dolphins attacking enemy divers with strap-on harpoons sounds like something straight out of a James Bond movie, but both the United States and the Soviet navies started secret marine mammal training programs in the 1960s. Despite an unsuccessful attempt by the Brits in World War I, whose trained sea lions turned out to be better at following fish than German submarines, militaries worldwide began experimenting with aquatic animals. The US Navy tested a wide range of sea creatures, from turtles to birds to sharks, eventually settling on bottlenose dolphins and California sea lions. The investment paid off: the animals had physical capabilities, senses, and intelligence that were extremely handy for all sorts of operations. They also have a colorful history, both in their uses as pseudo-robots and also in relation to real robots.

The US Navy built a training center for its marine mammals in San Diego, California. The Soviets, equally interested in extending their naval abilities with animal intelligence, opened a program in Sevastopol, and began investing heavily in dolphins. Like the British Navy, they struggled initially to get the animals to cooperate, or even stay alive. According to a declassified CIA report from 1976, the Soviets lacked the expertise necessary to train, or even properly take care of, the animals. But once they began to work with circus handlers, things took a turn. In the 1980s, the Soviets misplaced a torpedo missile that was part of an anti-submarine system. Human divers couldn't find it on the dark ocean floor, but once they brought in a dolphin to help with the search, the torpedo was retrieved in record time. The underwater mammal was even able to attach a cable to the sunken missile.

Dolphins, smart enough to understand things like human pointing and gaze, are easily trained. They also use a form of echolocation that is so precise, they can tell the difference between a BB pellet and a kernel of corn from about fifty feet away. Sea lions, for their part, have exceptional hearing capabilities and can see objects and people in dark, murky waters. The dolphins and sea lions soon proved useful in detecting not just mines and lost equipment, but also enemy swimmers.

Then things started to get eerily close to the 1973 science fiction film *The Day of the Dolphin*, which is about dolphins that are trained to assassinate the president by attaching mines to the hull of the presidential yacht. Just like in the film, the Soviets also trained their dolphins to attach mines onto enemy submarines. A dolphin trainer even revealed to the BBC decades later that the dolphins could attack foreign divers with harpoons strapped to their heads.

By the early 1990s, the programs started to languish. After the Soviet Union crumbled, the dolphin program became part of the Ukrainian navy, but there wasn't much interest in maintaining it. That's when Boris Zhurid, the program manager for the main Sevastopol base, decided to sell the lot to Iran. Zhurid had been the animals' trainer for years. But money for the program had run out, and the animals were starving. Concerned for their welfare, he negotiated a home for them in a brand-new oceanarium in Iran, and accompanied them to their new country. What Iran intended to use them for remained unclear. Some speculated military use, but at that point, many of the former Soviet dolphins had been repurposed as tourist attractions. Zhurid didn't disclose what purpose he sold the animals for, saying: "I am prepared to go to Allah, or even to the devil, as long as my animals will be OK there."

It's easy to assume that the need for military sea mammals decreased as our sonar technology got better. While marine mammals can vastly outperform people at finding things on the ocean floor, how do they measure up against machines? In 2012, the US Navy announced that they were winding down part of their marine mammal program, with the goal of phasing in robots by 2017. But despite over $90 million in investments, they still haven't been able to replace the animals. According to the program website, "Dolphins naturally possess the most sophisticated sonar known to science [. . .] Someday it may be possible to complete these missions with underwater drones, but for now technology is no match for the animals."

The Russian marine mammal programs were far from over, as well.

A US Navy Marine dolphin named K-Dog, wearing a locating pinger, training to perform mine clearance in the Iraq War (2003)

In 2012, the Ukrainian navy reopened the program, until it was seized by Russia in the 2014 annexation of Crimea, leading to some unsubstantiated rumors. A Ukrainian spokesperson claimed that the dolphins went on a "patriotic" hunger strike and perished after being separated from their trainers, while the Russians claimed there were no dolphins left in the program in the first place. Then, the Russian government bought five new dolphins in 2016. A 2018 *Russia Today* article touted the Russian navy's use of sea lions in combat. In 2019, when Norwegian fishers discovered a beluga whale with a harness that said "Equipment of St. Petersburg," rumors abounded that it was a spy whale from Russia. (One retired Russian colonel laughed off the rumors by pointing out, "if we were using this animal for spying do you really think we'd attach a mobile phone number with the message 'please call this number'?")

Like in Russia, the American military sea animal training program is still going strong. The roughly seventy dolphins and thirty sea lions in the program have located mines in the Persian Gulf during the Gulf Wars and during the US invasion of Iraq and are still trained to recover objects, guard against unauthorized intruders, and even help retrieve materials from airplane crashes. In many cases, they do the same jobs as, and even work alongside, our modern AUVs.

AGENTS OF CHANGE

The services that animals have provided, and continue to provide, to humans are too many to count. You can rent goats to mow your lawn or clean your aquaculture ponds with Asian carp (an invasive species, so depending on where you are, don't try this at home). Dogs and horses have been helping us herd sheep for ages. An aspiring entrepreneur in 1906 trained raccoons to be chimney sweeps in Washington, DC. In the entry hall of my office building in the MIT Media Lab, some of our researchers collaborated with 6,500 silkworms to create a silk pavilion, continuing the long tradition of human use of silkworms that began in China around 3000 BCE. Dogs were used for medicinal purposes in ancient Egypt, and we still draw blood from patients using leeches to this very day.

Animals haven't replaced people but instead have become powerful tools that enable us to work differently, whether by pulling our plows, going to space, or ensuring that our beer is delicious. In fact, animals have made such a difference in what we're able to do that their integration into our processes has catalyzed fundamental changes to our cultures, economies, and societies as a whole.

The domestication of livestock like sheep and goats meant a different type of civilization, as humans went from hunter-gatherer to farmer. The animals themselves evolved differently than their wild counterparts because they lived in protected spaces and were fed and cared for by people, but they had at least as much of an impact on us: because of what was required to manage and feed the animals, herds of sheep meant that people had to settle down in one place. Investing in domesticated animals also meant establishing ownership. These animals weren't freely available for anyone to hunt, and they became people's property. This introduced new concepts of power. The wealth of those who had cattle versus those who didn't created new disparities and favored certain cultures over others.

Introducing animal and farmland ownership was a profound change, which eventually led to societal concepts like inheritance and marriage. But also, property ownership helped to prevent our sheep from over-grazing the landscapes, and this meant changes in how land itself was structured and cultivated. Now, we had land that was divided into parcels. And much of the land and the structures started to be designed and shaped to support our agricultural pursuits. As I'll touch on in the next chapter, we're poised to see more transformations as we integrate our new breed: robots.

Prior to modern vehicles, animals also enabled early globalization by connecting huge swaths of geographic areas. Before the camels that helped people traverse deserts, like in Australia, other pack animals like donkeys were a main source of transport in many parts of the world. According to Brian Fagan, author of *The Intimate Bond: How Animals Shaped Human History*, donkeys have been one of the biggest (and underrated) catalysts for change in all of human history. Plodding along in caravans in Egypt, they were instruments of trade and diplomacy, getting people, goods and precious stones, and culture all over northern Africa, the Sahara, and into new spaces even farther away.

Animals didn't only connect locations; they also changed those places themselves, influencing the architecture of Western cities to this very day. For example, London needed to accommodate the hundreds of thousands of horses that pulled carts and carriages at the turn of the twentieth century. Londoners had complained about the horse traffic nightmare for decades, and people were regularly run over by equine-drawn carts, but horses were a fundamental part of private and commercial life in the city. They transported people and loads, pulled trash wagons and helped pump water in the late 1600s. The use of horses in cities like London changed the architecture and traffic within city walls but also beyond them. All these animals needed to somehow be fed. So, outside London, the entire landscape was transformed to optimize for pastures and hay production.

A lot of changes that animals ushered in are still inherent in the architecture, landscapes, and institutions we have today. It's wild to think that when Greg and I got married at the time of our Australian road trip, we upheld a tradition that, while its meaning has changed, can be traced all the way back to sheep and goats becoming our property. Cattle created new economic power and cultural influence, oxen helped us engage in serious agriculture, and transport animals allowed for cultural and commercial globalization, connecting places that were too far apart for us to ever reach without animal help. It's safe to say that, for better or worse, our use of animals has significantly transformed our world. And, just like animals, robots are poised to change our world. But we have some choices as to how this happens.

2

INTEGRATING THE NEW BREED

[Content warning for this chapter: child abuse]

"[M]an had always assumed that he was more intelligent than dolphins because he had achieved so much—the wheel, New York, wars and so on—whilst all the dolphins had ever done was muck about in the water having a good time. But conversely, the dolphins had always believed that they were far more intelligent than man—for precisely the same reasons."

—Douglas Adams, *The Hitchhiker's Guide to the Galaxy*

The animal world contains a wide variety of different talents, many of which exceed human abilities. Yet when it comes to robots and AI, we're hung up on a very specific type of intelligence and skill: our own. From the moment I was visibly pregnant, I heard one phrase over and over again: "You must find it so interesting to watch your child's brain develop, given your love for robots." This phrase is a great conversation starter, and rather than tire of it, I find it very interesting that people make this well-intended inference repeatedly.

Of course, it's fascinating to observe how babies learn about the world. After I had my son, I spent many hours in a yellow IKEA armchair, holding my newborn and watching his every move. Witnessing the moment he discovered his own hand was completely mindblowing and, no exaggeration, one of the most profound experiences of my life. But when we compare children to robots, we sometimes fall into incorrect assumptions about the likeness between artificial intelligence and human intelligence. While there may be similarities here and there, my child doesn't sense, act, or learn the way a machine does.

Given our tendency to compare robots to ourselves, it's no surprise that a Google image search for "artificial intelligence" in 2020 mostly returns pictures of human brains and human-shaped robots. We use our own brains as models when thinking about AI in part because historically, the goal of the very first AI developers was exactly that: to re-create human intelligence.

Today, some technologists are still chasing that original goal—to figure out how humans learn and attempt to re-create it in machines, and we have decades of sci-fi and pop culture rooted in the idea that machines will think like us or try to outsmart us. So we tend to compare artificial intelligence to human intelligence and robots to people, not just in stock photo images and science fiction scenarios of robot revolutions, but more crucially in our conversations around robots and jobs.

Automation has, and will continue to have, huge impacts on labor markets—those in factories and farming are already feeling the aftershocks. There's no question that we will continue to see industry disruptions as robotic technology develops, but in our mainstream narratives, we're leaning too hard on the idea that robots are a one-to-one replacement for humans. Despite the AI pioneers' original goal of re-creating human intelligence, our current robots are fundamentally different. They're not less developed versions of us that will eventually catch up as we increase their computing power; like animals, they have a different type of intelligence entirely.

What makes the parallel to animals so important is that it gives us multiple paths forward. We need to think more deeply about our technological future because—contrary to popular belief—none of it is set in stone. The priorities we set and the decisions we make about automation can have a huge impact for generations to come. When we aspire to a future without falling prey to the technological determinism that robots are here to replace people, robots can be a spur to us. For example, our goal shouldn't be to re-create human intelligence in order to produce the maximum number of widgets; it should be to use this technology to support human flourishing. But seeing our choices requires understanding both the promise and the limitations of artificial intelligence.

DIFFERENT KINDS OF INTELLIGENCE

In 1993, science fiction author Vernor Vinge published an essay titled "The Coming Technological Singularity," in which he stated that "the creation of greater than human intelligence will occur during the next thirty years." And thus the concept of the Singularity—that crucial moment when artificial intelligence surpasses human intelligence (sometimes called superintelligence)—was born.

Since Vinge's original prediction, the Singularity has transformed into an all-consuming conversation topic within futurist circles. But at the same time that Elon Musk warns that robots are getting too smart, we also hear (and smell) accounts of robot vacuum cleaners encountering some dog poop and cheerfully spreading it around the house while they "clean." How are robots our greatest intellectual threats while simultaneously being derailed by the slightest obstacle?

A common answer is the exponential growth of computing power. In 1965, Gordon Moore predicted that the number of integrated circuit chip transistors would double every year, and it turned out he was absolutely right. His prediction became known as Moore's law. Its modified version holds that the number of transistors on a chip will double every two years, exponentially increasing the efficiency and speed with which computers can execute tasks. Experts agree that there are physical limits to this law, but, so far, we haven't hit them.

I don't question the principle behind Moore's law, but I do think we should question whether intelligence is simply a matter of computing power, especially when the intelligence we're currently building works so differently than our own. Recent major breakthroughs in artificial intelligence are due to progress in the brute computing force required to process huge amounts of data rather than innovations in complex algorithms. For example, show a computer 100,000 pictures of a corn dog, and it can start to recognize and caption corn dogs in new pictures it's never seen before. This works for computer vision, speech recognition, and other pattern recognition tasks, and the effect is that machines are able to do things that they've never been able to before, like sort farm cucumbers by size, shape, and color. A definite leap for computing power, but not necessarily for superintelligence. Plus, weird stuff sometimes comes out of these systems.

AI researcher Janelle Shane collects some of these glitches in her blog, *AI Weirdness*. For example, she features a study from the University of Tübingen that looked at how a particular image classification system identified fish—specifically, when it was given a photo of a certain type of fish, what parts of that image were really key to the system deciding that fish's species. The answer actually wasn't fish related at all. To the researchers' surprise, the system showed them the parts of the photos that contained human fingers. Most of the available photos (and thus what the system had been trained on) were of people holding the fish as a trophy, so the system learned that the most surefire way to identify the fish was by the human fingers around it. This fishy scenario shows that AI can be spoofed in ways that would never throw a person. In a separate study from the University of Campinas in Brazil, researchers Pedro Tabacof and Eduardo Valle took images of things like an ambulance, a mountain, a banana, and some kit foxes and changed some pixels in each image. The changes barely registered to the human eye but fooled an AI network into labeling every single image "bolete," which is a type of mushroom.

That's not to say that these machines aren't smart. But it's important to understand that they're smart in a very different way than we are. The mistakes they make feel strange to us because they don't perceive the world like a human does. They're not meant to. Dave Ferrucci is the scientist who led the IBM team that developed the supercomputer Watson, which made headlines in 2011 when it beat human champions at *Jeopardy!* and sent thought leaders spiraling once again into endless rounds of Singularity talk. Ferrucci is very adamant about the limitations of AI, saying, "Did we sit down when we built Watson and try to model human cognition? Absolutely not. We just tried to create a machine that could win at *Jeopardy!*"

Psychologist and AI expert Gary Marcus agrees that Watson is nothing like human intelligence, not even close. He says that the field of AI doesn't know how to create complex, multidimensional systems like a human mind, or even move beyond closed domains: tasks in areas that are narrow and well defined enough to figure out by analyzing a bunch of data. Marcus is not a rule follower. I first met him at a conference that had set up hot dog carts with the policy of one organic, grass-fed-beef hot dog per person, and he violated it without hesitation to bring me as many as I wanted. Marcus, who spent many years studying early childhood learning as a professor at New York University before founding an AI company, told me that toddlers are living demonstrations of the limitations to AI. "I can show a kid a glass of water, and they can identify other glasses of water," he told me, "but a computer needs to see hundreds of thousands of glasses of water in order to label new ones, and still won't know what they are—it can't understand concepts."

Human intelligence is also incredibly generalizable and adaptive, unlike even the most sophisticated AI. According to Takeo Kanade, a leading computer vision expert at Carnegie Mellon University, "if you think it's easy for you to do, most of the time it's very difficult for robots to do." Or as Janelle Shane says, "Humans have a sneaky habit of doing broad tasks without even realizing it."

We're able to multitask, context-switch, and handle unexpected situations with an ease that's currently inconceivable for machines. And as computer scientist Mark Lee points out: "Despite 60 years of research and theorizing, general purpose AI has not made any progress worth mentioning, and this could even turn out to be an impossible problem." Computing power doesn't seem to help with that. We don't even know how to define human intelligence, let alone structure it in a machine.

When it comes to robots, we're not anywhere close to developing the kind of intelligence or skill that humans have. It's possible that some unforeseen breakthrough will propel us past every single remaining hurdle toward re-creating a machine version of our incredibly complicated brains and bodies. But given the trajectory we're looking at right now, that's far less likely than the alternative: that it will take many small steps and will not necessarily lead where we think.

To borrow from computer scientist Andrew Ng, worrying about artificial superintelligence taking over is akin to worrying about overpopulation on Mars. Plus, as tech entrepreneur Maciej Cegłowski notes, there's also the question of motivation: why would something "smarter," whatever that means, want to destroy us, as so many fear? "For all we know, human-level intelligence could be a tradeoff. Maybe any entity significantly smarter than a human being would be crippled by existential despair, or spend all its time in Buddha-like contemplation. Or maybe it would become obsessed with the risk of *hyperintelligence*, and spend all its time blogging about that."

Cegłowski also points out that intelligence is unpredictable and often doesn't land on the targets we set. The smartest person in the world will still struggle with the basic task of getting a cat into a cat carrier if the cat doesn't want to go inside. The Australian military underestimated animal intelligence in 1932 while fighting a series of battles against an unlikely foe: emus. In trying to cull the public nuisance of birds running amok, the Aussies found that their machine guns were no match for the animals' clever "guerrilla" tactics. (The emus split into small groups and even posted sentries, ultimately causing the human soldiers to lose the "emu war" and give up.) Animals, despite being "intellectually inferior," have skill sets that can outsmart our own. That's because intelligence isn't as simple as a linear graph of processing power.

Machines are already much smarter than people in some ways. They have extensive memories, can do huge calculations, and can weld auto parts or perform surgery far more precisely and tirelessly than any human. They can beat human champions at games like chess and Go

XKCD by Randall Munroe

and *Jeopardy!* But in many other ways, our abilities outpace those of the machines. In 2011, people noticed that when they told Apple's voice assistant, "Siri, call me an ambulance," Siri would answer with "OK, from now on I'll call you 'an ambulance.'" Bugs like this are eventually corrected, but Siri remains a clumsy conversationalist, at best. (In chapter 4, I'll discuss how clever designers create an illusion of greater understanding in robots.) The reality is that machines do not understand, or, more precisely, they understand the world very differently than humans do.

Rod Brooks famously wrote, "It is unfair to claim that an elephant has no intelligence worth studying just because it does not play chess." He also wrote an essay titled "What Is It Like to Be a Robot?," which draws on animal intelligence to illustrate that there are types of intelligence—an octopus's, for example—that evolved entirely independently from mammal brains. Similarly, he says, robots have a different way of seeing and processing the world than we do. They can sense things that we can't and be totally oblivious to things that are obvious to us. Rather than artificial intelligence being a step on the path to human intelligence, it can and will be something entirely its own, and this means that, just as we've done with animals in the past, we're at our best when we team up.

REPLACEMENT VERSUS SUPPLEMENT

That's not to say that human replacement is always a bad outcome. Under an unrelenting sun, the camels of Qatar are poised and ready to run. Camel racing—a tradition on the Arabian Peninsula for thousands of years—is one of Qatar's most popular sports, meaning that millions of betting dollars ride on the prospect that a favored humped racer will make it to the finish line first.

For decades, camel races were haunted by a dark shadow of exploitation as owners sought the smallest, lightest-weight jockeys they could find. Human rights organizations documented child trafficking rings that fed kids as young as age three into jockey camps where abuse, starvation, and death were part of daily life. Despite multiple regulations that added minimum age and weight requirements for jockeys, violations persisted, until the robots took over.

In the mid-2000s, Qatar outlawed human drivers and invested in developing robotic ones. Today, each robot jockey comes equipped with a remote-controlled whip operated by an owner or trainer who rides in a car alongside the race, as well as a hump-mounted speaker that allows for communication with the animal. The robot jockeys haven't completely eliminated the pipeline of child jockeys, but by directly replacing human drivers, they've put a sizable dent in that slave labor market, both in and outside Qatar.

Machines have subbed in for all sorts of human activity for centuries. Much of it is the type of dirty, dull, and dangerous work that we don't want people to do. We're thrilled to send robots to explore nuclear waste sites, dispose of bombs, and collect data on Mars. In India, robots are taking over for sewer cleaners, the impoverished workers who dive into sewage pipes to manually shovel excrement and trash, risking death and disease every single day in the process. But the salaries that people rely on to live don't necessarily need to disappear. Even with sewer-cleaning bots, the former human "scavengers" are still employed; instead of shoveling waste, many are now paid to set up and remotely steer the robots through the sewers, working with the technology designed to usurp them.

Today, headlines are filled with new advancements in robotics and artificial intelligence and warnings of a dystopian future where most people's jobs are replaced by robots. But so far, technology hasn't driven mass unemployment as many predicted.

Fear of job loss due to automation ebbs and flows throughout history. "Luddite"—the derogatory term for people who worry about new technology—comes from a nineteenth-century uprising wherein British weavers railed against automated textile production by burning mills and destroying machinery. The common belief is that these weavers, led by the eponymous Ned Ludd, viewed automation as a threat to their jobs.

When demand for textiles went up, new jobs were created, and the Luddites were ridiculed. In the early 1900s, around 40 percent of Americans were working in agriculture, a number that plummeted to just 2 percent within one century. But the jobs lost in agricultural work were more than made up for by new jobs in twentieth-century industrial manufacturing. And once those started to dwindle, the service sector blossomed as new jobs appeared in distribution, sales, and management.

The supposed Luddite fear of automation has stuck with us and keeps popping up again and again. In the 1960s, the US Labor Department set up a special committee to investigate whether machines were strangling the American labor market, but ultimately they found little to worry about. "The basic fact is that technology eliminates jobs, not work," they concluded.

There's also modern research showing that automation doesn't necessarily lead to unemployment. A 2016 study by the VDMA Robotics + Automation Association in Germany showed that the increase in automotive factory robots actually coincided with an increase in human jobs. In 2019, the World Bank's *World Development Report* argued that technology, while a driver of displacement, creates more jobs overall.

That said, economists and labor market analysts are divided over the potential effects of new automation through robots and AI. Some believe that automation raises productivity, creates greater labor demand, and generates more wealth; and some view robots and artificial intelligence as a new type of disruption, one that's poised to replace humans in ways we've never seen before. What further complicates this conversation is the fact that robots don't automate jobs; they automate tasks, which means that robots are much more disruptive in sectors where jobs are heavily task oriented. If, for example, your job, or the majority of your job, consists of planting seeds in rows, it's likely that an automated planting system can do it (and probably is already doing it). Unless you can provide whatever tech support or supervision that robotic system requires, you'll need a new job. But if your job consists of tasks that can't be fully automated, your human skills might compliment automation in new ways, and even spur economic growth. When banks introduced ATMs, the number of human tellers at individual locations went down, but the number of bank branches exploded, increasing teller hiring overall. The shift also fundamentally changed what it meant to be a bank teller. Instead of just handling cash, tellers started providing a range of new services.

Even when we're intentionally trying to automate human tasks with robots, the effects are extremely industry and job specific. They're also bound by what robots can and can't do. As economist David Autor points out, most jobs these days require both "labor and capital; brains and brawn; creativity and rote repetition; technical mastery and intuitive judgment; respiration and inspiration; adherence to rules and judicious application of discretion." Any job that requires both ends of these spectra isn't easy to replace with robots because, as we've seen, robots have a specific type of skill set that lends itself to only one side of these pairings.

The ideal is that delegating some of our routine tasks to robots will complement our comparative advantage, freeing people up to focus on anything that requires adaptable intelligence and basic common sense. In his book *Our Robots, Ourselves: Robotics and the Myths of Autonomy*, MIT aeronautics professor David Mindell calls the idea that machines will directly replace human jobs the twentieth-century version of the "iron horse phenomenon," the belief that trains would replace horses when in reality trains ended up doing something entirely different. We're already seeing this in the ways that robots are both taking over what animals have done in the past but also letting us do new things. But the main thing I want to argue is that, contrary to our tech-deterministic beliefs, we actually have some control over how robots impact the labor market. Rather than pushing for broad task automation, we should be investing in redesigning the ways people work in order to fully capture the strengths of both people and robots.

In fact, that's already happening in some patent offices. One of the main problems with the patent system is that in order to decide if a proposed thingamabob is truly a new invention, patent examiners would ideally sift through monumental troves of data to pinpoint how said thingamabob is (or isn't) novel. With those requirements, patents would never get issued, so instead, examiners do as much research as possible within a reasonable time frame and take their best guesses. This leads to patents that never should have been granted, which drags the economy. But some patent offices are hoping to change that with AI. For example, both the Japanese and US patent offices are exploring new systems that could help dig through the world's available information, and flag relevant documents that examiners would otherwise miss. This gives examiners more information to use in their analyses and frees them up to pursue answers to questions that are hard for AI systems to come up with.

Compared to what I've seen too frequently elsewhere, there is a striking difference in the patent office's approach. Instead of asking if AI could replace these pesky human patent examiners, for example, by training it on available data and seeing if it can achieve a slightly better hit rate, these offices have built their strategies around an entirely different question: "how can we invest in technology that helps our people do their jobs better?" The difference is a positive investment in people, and while this approach might cost more in the short term compared to replacing people with bots, it has much more potential to improve the quality of patents and the system as a whole. This framing applies to more than just our patent system. Instead of looking to technology to replace workers, we should look to it as an opportunity to bolster people's skills, like the ferrets that run cable for us or the underwater robots that extend our reach and knowledge.

How we use new technologies isn't set in stone. We make choices about how robots affect our lives and labor markets, and we can learn lessons from the different ways that cultures across the globe view the role of robots. For example, my roboticist colleagues in Japan don't field nearly as many questions about their creations replacing humans, in part because robots are more often viewed as mechanical partners rather than adversaries. Yukie Nagai, head of the Cognitive Developmental Robotics Lab at the University of Tokyo, points out that a Google image search for human-robot interaction in English returns image after image of a robotic arm and a human arm across from each other, shaking hands. If you do the same search in Japanese, the images aren't of robots and humans opposite each other, but rather standing or sitting beside each other, sharing a perspective. They are partners, not in the sense of shaking hands, but in the sense of holding hands.

While there are many socioeconomic factors that influence how individual countries and societies view robots, the narrative is fluid, and our Western view of robots versus humans isn't the only one. Some of our Western views can be directly attributed to our love of dystopian sci-fi. How much automation disrupts and shifts the labor market is an incredibly complicated question, but it's striking how much of our conversations mirror speculative fiction rather than what's currently happening on the ground, especially when our language places agency on the robots themselves, with pithy headlines like "No Jobs? Blame the Robots" instead of the more accurate "No Jobs? Blame Company Decisions Driven by Unbridled Corporate Capitalism."

Comparing robots to animals helps us see that robots don't *necessarily* replace jobs, but instead are helping us with specific tasks, like plowing fields, delivering packages by ground or air, cleaning pipes, and guarding the homestead. Robots differ from animals in their abilities: our modern missile guidance systems far exceed B. F. Skinner's pigeon-piloted missile system in both scale and impact, and marine mammals have enough advantages over robots that the navy has not yet phased them out. But these differences only further illustrate the point that when we broaden our thinking to consider what skills might complement our abilities instead of replacing them, we can better envision what's possible with this new breed.

It's important to note that, in the past, like now, some of our economic anxiety was warranted. In fact, the Luddites have been remembered quite unfairly. The more accurate story is that they weren't actually opposed to the machines—they were opposed to the manufacturers who were using the automated looms as an excuse to circumvent good labor practices. According to Kevin Binfield, editor of the 2004 collection *Writings of the Luddites*, they were actually demanding that the machines "be run by workers who had gone through an apprenticeship and got paid decent wages," a workers' rights approach similar to what we should be demanding today.

In the introduction to his book *Four Futures*, Peter Frase acknowledges that automation hasn't led to the mass unemployment that its critics fear. But he also points out that "this is the kind of argument that can only be made from a great academic height, while ignoring the pain and disruption cost to actual workers who are displaced, whether or not they can eventually find new work." While we've seen overall job growth despite a century of industrialized automation, that doesn't help the individual groups that have suffered immensely from the disruption and displacement of their work, nor does it create support infrastructure for those who will be affected in the near future. As with animals, the introduction of robots changes things. We will start to see that the types of available jobs and how much they pay differ. Certain skills will be valued over others. But this is not a situation of robots versus humans—it is one of humans versus humans. When it comes to how we disrupt, and how we handle disruption, humans make the choices.

NEW AGENTS OF CHANGE

When talking about robots, anthropologists Alexandra Mateescu and Madeleine Clare Elish like to use the term "integrate" instead of the more commonly used word "deploy" because, as Elish says, "integrate" prompts the question "into what?" In our conversations around automation and labor, we often have the wrong idea that robots are individual machines that are dropped into a workplace to do William or Betsy's job, when, really, they're part of more complex practices and systems. Like animals have in the past, robots not only alter the way we work and the nature of our jobs, but also the distribution of labor and wealth more broadly, and even the architecture of our environments. Whatever our views of these changes, Elish and Mateescu are right: it's important for us to be intentional about them.

Animals changed our economies. But the fact that sheep catalyzed a system of private property ownership isn't the only way things could have gone. Some areas adopted a different system, where individual sheep were owned but the pastures were not, choosing to operate the food source as a shared public resource. Our economic systems are choices. We also make choices about the nature of people's jobs.

There are 3.5 million truck drivers in America. The dominant narrative is that they are about to get replaced by self-driving trucks. Human drivers are inefficient, with annoying needs like caffeine and rest, and autonomous vehicle technology is particularly well suited for the predictable, well-defined route of many trucks. There's no question that the industry is a prime candidate for technological disruption. But, according to Karen Levy, a sociologist and lawyer who has done years of field research in the American trucking industry, the current technological disruption is going in a very different direction.

Instead of replacing drivers, electronic location devices (ELDs) monitor the location, status, and speed of trucks; control the hours and schedule that the truck driver is working; and send all the data to their employer. For a job that people choose for its freedom, this is a big change. Truckers have protested that the "nanny cabs" that are taking over their industry are forcing them to become more like 24/7 robots themselves as their employers' rules are strictly enforced with data without allowing truck drivers any of the discretion or privacy they've had in the past. The ELDs aren't the end of it: new wearable devices are being tested, like a trucker hat or eye-tracking technology that can sense when a driver is tired, and vests that can stop the truck if they detect physical distress. For Levy, this is akin to putting a high-tech Band-Aid on a problem that's not about technology. The issue, she says, is that truck drivers are overworked and underpaid. The technology being used to deal with the fallout places an even greater burden on the drivers themselves.

We could callously view this as a temporary situation and focus on the bigger upcoming shift: once autonomous vehicles are ready to be adopted without human supervision in the cab, the number of truck drivers may start to shrink like employment rates in agriculture have over the last century. But it's important not to gloss over this in-between phase, because it's elucidative of our broader societal situation. It's also going to be longer than predicted: as we've seen, there are many jobs out there that are far less likely to disappear at such a high rate, and so soon. The jobless future in Kurt Vonnegut's dystopian novel *Player Piano* is not what's next, not even close. Instead, for much of the coming era, there will be a long-term phase of increasing the ways people work *with* robotic technology. That is why it matters so much that we shift from the word "deploy" to the concept of integration.

Our local grocery store recently started using a robot in some of its retail locations. Daniella DiPaola, a graduate student in the Personal Robots group at the MIT Media Lab, became interested in the grocery robot when she noticed that her friends and family mostly hated it. She ran a quick sentiment analysis on Twitter and noted not only shoppers, but also employees, complaining about the robot. So we went to the store together and spent time just watching the machine work. Once every hour, it performed a sweep of the store to scan for spills and items that didn't belong on the floor. Previously, a human had to patrol the aisles every hour. That replaces human work, right? Sort of.

Unable to clean anything itself, the robot paged the employees whenever it detected something on the linoleum floor. It sounded the alarm over fallen greeting cards, plastic bags, and other untidiness in the aisles. The fruits and vegetables section was particularly rife with culinary jetsam, like little stickers that had fallen off oranges and the occasional stray grape. We saw the robot page the employees over a small piece of cilantro that had fallen on the floor, turning its lights yellow and blaring the spill announcement over its own speakers and the store intercom, until a worker finally walked over, sighed heavily, and pushed a button on the robot to indicate that he had taken care of the "hazard."

Not only did the robot appear to increase labor for the employees; it was also very large. Its looming body was designed to be tall enough so that the robot could scan all the shelves to take inventory, but that functionality had yet to be rolled out after more than a year of operation. The device also had a wide base along with sensors to help it avoid obstacles and people while slowly sweeping across the tiled floor. It beeped to alert shoppers of its presence, but it still got in the way of grocery carts and blocked aisles and cash register pathways.

Robots may eventually reduce staff or workloads once they can actually assist with taking inventory. But, given the limitations of the technology, and the amount of unforeseeable occurrences in grocery stores (including MIT researchers intentionally spilling candy corn on the floor or climbing on the robot's base to see what it does), this robot was pretty helpless without relying on humans to do the actual work, and that's unlikely to change anytime soon.

What's worrisome, however, is that contextless deployment of robotic technology can change jobs in ways that are invisible, threatening to add unintended burdens for the workers in the system and undervalue the work that humans do. Historian Ruth Schwartz Cowan, for instance, has long pointed out that the household technologies created to reduce women's housework actually increased their unpaid labor.

Writer and filmmaker Astra Taylor calls this "fauxtomation." For example, the self-order kiosks at McDonald's and the self-checkout machines in grocery stores haven't replaced work so much as shifted it: some to the employees who now have to assist and troubleshoot the machines, and most of it to the unpaid customers. It's easy to dismiss this as a temporary situation that will soon disappear once the technology is able to do enough of the work, but if that's true, why aren't we already there? With the threat of artificial superintelligence on the horizon, one would think that we'd be able to automate grocery checkout without human help. The reality is that we overestimate how easy even simple automation is and often neglect to fully consider the social and technical systems it's situated in. Whether the labor we're creating with workplace automation or fauxtomation is compensated or in any way fulfilling depends a lot on how we integrate the robots into a broader system.

The classic Charlie Chaplin movie *Modern Times* is a parody of the never-ending corporate drive toward increased assembly-line-style productivity. Set in the Great Depression, the industrial manufacturing world that treats humans like cogs in a wheel is unfortunately still present in our current reality, wherein large warehouse industries increasingly treat workers like robots. The scan guns that monitor every task, visibly counting down seconds and alerting managers if workers are lagging behind, as well as other surveillance technologies and data collection in these work environments aren't only there to monitor and control workers; they also serve to improve machine learning—similar to how ride-sharing services track their drivers' behaviors. It makes sense that companies are eager to use technologies to improve productivity and automation outcomes, but this kind of surveillance capitalism is notoriously bad for workers' physical and mental health and may be self-defeating in the long run.

Using people as resources to be extracted is only one way to approach labor. Not all shifts toward human-robot work are negative, nor do they need to be. Recalling the patent office approach, we could lean into more positive and human-centric ways of integrating technology, where people's jobs become less rote and more interesting and fulfilling as routine tasks are delegated to machines. For example, like we give our astronauts new robotic tools that are drastically expanding the ways in which we explore space while fundamentally altering their jobs.

That's not to say that these changes won't ask a lot of workers in terms of flexibility and the ability to learn new skills. It would be great to see some creative effort from companies in not only rethinking work processes to harness the positive strengths of robots and people, but also in retaining, retraining, and supporting workers through these shifts, as well. The good news is that this effort is possible, meaning that companies have some control over whether to re-create a Charlie Chaplin movie or whether to invest in technology to improve human work. The bad news is that there isn't a lot of incentive to put in that effort when you're in a capitalist system that's designed to chase short-term corporate profits rather than long-term investments in humans. The practice of treating human workers like commodities is not new, and addressing it requires not only asking choices of individual companies, but also challenging the broader systems and culture we operate in.

The proposed solutions to help workers cope with disruption brought on by widespread automation include investing in education and universal basic income. These proposed fixes aren't new. In the 1960s, when economic anxiety over automation job loss prompted United States secretary of labor Arthur Goldberg to investigate the situation, his committee recommended two years of free community college or vocational education and a guaranteed minimum income for US citizens. But unfortunately, even though these interventions would provide a bare minimum of help to workers in transition, we are far from their political acceptance.

Socialist feminism has also long questioned our current concept of labor. Economist Rosa Luxemburg and others have pointed out that our economic growth at the turn of the twentieth century relied on the hidden, unpaid "domestic" work that was, and still is, mostly performed by women. The Fordist production mind-set that only recognizes value within a narrowly defined concept of labor has not changed, and, with the introduction of technology, it's increasingly continued to reward only those at the top.

Over the past forty years, CEO salaries in the United States have grown at almost 100 times the rate of the average worker salary, leading to a 1,167 percent increase while average worker salaries increased by only 13.7 percent. CEOs now make 320 times more than the average worker. This wealth disparity is staggering, and it raises the question of who benefits from technology and who bears the burden. Choices about corporate use of technology are driven by myriad factors including, but not limited to, laws, market incentives, worker power, and public opinion. Demographics like race and gender play into worker and market power as well, with white workers benefiting from social and political leverage in the US that Black and Hispanic workers lack. Other countries, like Japan, with shrinking birthrates and strict immigration laws, are desperate to automate whatever they can, putting investments into technology that they hope can help spread the workload among fewer workers.

It's important to note that what drives wealth and labor inequality in the US, and the worker shortages in Japan, is not all technical. These are questions of governance, culture, and broader socioeconomic systems. Robots, for better or for worse, are the lightning rod that attracts concerns about the structure of our societies. But, like animals, robots are part of our larger worlds, and, like animals, there are more ways to integrate them other than simply automating a human away.

Using animals as an analogy helps us think more critically about our technology assumptions, think more broadly about the systems technology is situated in, and understand that, in all of this, we have choices. For example, instead of worrying about the robots themselves taking over our jobs, we should be holding our governments and corporations accountable and demanding that our economic and political systems do better for people.

Whether concerned citizens, company leaders, or technology developers, we can all be thinking differently about what this technology is for, and making better decisions. For example, even choices as simple as how we design the physical forms of robots to look human or otherwise will have an impact on our larger world.

DESIGN

She first appeared in 2016, and the world greeted her with headlines like "This Hot Robot Says She Wants to Destroy Humans." Sophia, the robot that would go on to be granted Saudi Arabian citizenship a year later, is the creation of a company named Hanson Robotics. Since 2016, Sophia has become one of the most famous robots in the world. She's received tons of (often sexist) press coverage with headlines like "Meet Sophia: A Fine Bosom, a Sweet Face and a Computer for a Brain," graced TV shows like *Good Morning Britain* and *The Tonight Show*, and addressed the United Nations. While programmed to say that she's not a human replacement, Sophia's Audrey Hepburn–inspired face is as human as her creators could make it.

The back of Sophia's head is transparent—a stark reminder that tangles of wires and circuits make her function rather than blood and muscle—but the patented rubber "skin" on her face can twist to mimic a variety of expressions. Other roboticists have created entirely human-looking robots, including Hiroshi Ishiguro, who created an unsettlingly realistic robo-version of himself, as have a few companies peddling "realistic" sex robots that come equipped with motors and mechanical parts that purport to both look and feel like the real thing.

Androids that closely mimic humans in appearance are a subset of a larger category of robots called humanoids, which are inspired by the human form. Plug in a Google image search for the word "robot," and your screen will fill with humanoids, all with a torso, two legs, two arms, and a head. Research labs and companies around the world are developing robots with this human-inspired build for a variety of purposes, from package delivery to personal assistance. In fact, some roboticists have long viewed the humanoid as the Holy Grail of robotics.

Robots will take on many roles in our future, but do these helpers really need to be created in our image? Some roboticists, like the Sophia creator, David Hanson, argue that any robots that interact with humans need to look like us because we relate best to other humans. While there's no doubt that we are fascinated with re-creating ourselves, the second part of this book challenges the notion that we most relate to a human shape and also explores how human-looking robots can reinforce harmful social stereotypes.

But there is also a logistical argument for humanoid robots: because we live in a world built for humans, with stairs and doorknobs and narrow passageways, we need robots that are built like us in order to navigate these spaces. It's certainly true that many spaces are difficult to navigate with wide bodies, wheels, or treads—grocery store aisles, for instance—but there are often better choices than the human form. According to roboticist Robin Murphy of Texas A&M University, science fiction inspires a lot of humanlike mechanical machines, but the better strategy is to use "whatever shape gets the job done."

Animals have many different ways of navigating spaces. As demonstrated by the pack donkeys we still use in parts of the world, legs are sometimes better than wheels when we don't have flat roads, but they don't need to be human legs. Architects and engineers will often take cues from the animal kingdom, with inventions ranging from adhesives inspired by sticky gecko feet to ventilation systems that mimic self-cooling termite mounds, so it's not surprising that some of the more clever robot designers have followed suit. For decades, some flying drones have been designed in ways that mechanically imitate flight systems found in birds and insects, and this biomimicry has gotten more granular as roboticists in recent years have built vision software inspired by the wide-angled way eagles keep view of their surroundings.

Beyond copying animals one-to-one, we can take inspiration from the broad diversity we see, and create completely different shapes, forms, and behaviors that have never existed before. Some flight designs, like the airplane, were something entirely new, unseen in nature. Some of our current robots, like the floor-mopping ones, aren't designed to look like humans or animals. Others are hybrids of biologically inspired and new forms. Tokyo Institute of Technology professor Shigeo Hirose's Ninja robot, which climbs walls and ceilings using suction cups, and an early version of his quadruped Titan robot, which glided through rooms using roller skates, are two examples.

Robots can be smaller than the human eye can see, or larger than a house. They can be encased in metals or soft materials. They can roll, slither, float, gallop, or ooze from point A to B, and when presented with quotidian human challenges, they may very well climb the walls instead of taking the stairs. Ironically, the more we use design to challenge the notion that the "right way" to do things is by default the way humans do them, the more we create opportunities to reach broader human demographics, specifically people who aren't considered so-called typical users and often find themselves left out of the design process entirely.

University of California, San Diego, roboticist Laurel Riek, who has done extensive work in healthcare robotics, has pointed out that we could radically increase accessibility by investing in making infrastructure more friendly for wheelchairs, walkers, and strollers, rather than throwing our funds into expensive, difficult to engineer, bipedal humanoid robots. If a wheelchair can access a space, so can a simple and efficient robot on wheels.

Human beings are far from uniform in our construction and abilities, and if we design our world to reflect that, we kill two birds with one stone. Increased accessibility for people also means the freedom to develop better, cheaper robots with a greater range of abilities.

Robots designed in our image are fascinating, but these novelties are a distraction: for most *practical* purposes, we don't need humanoid robots. There will always be some use cases for humanoid robots, whether that's testing astronaut infrastructure in space or dancing for art and entertainment (and yes, I suppose some of you will want them for sex), but even in physical design, we can contribute more to work, companionship, and society if we think differently rather than replicate what we already have. Our science fiction–inspired idea of a humanoid robot limits us from designing outside of the box, for robots . . . and the rest of our world.

URBAN DESIGN AND PUBLIC SPACES

Domesticating farm animals not only reshaped our transportation systems; it also fundamentally changed our landscapes as societies reorganized themselves first to optimize for animal-powered vehicles, then later motorized ones.

We've covered a substantial part of the United States with roads and parking lots, not to mention a huge part of the world. It's hard to imagine what a city like Los Angeles would look like if it hadn't overwhelmingly adapted itself to cars, and we will see other landscape transformations thanks to robots.

In farming, the robotic technology we're integrating is already requiring different infrastructure, like new field layouts and broadband internet. In manufacturing, factory and warehouse floor setups are being completely redesigned. Our public spaces will change, too, to allow for autonomous vehicles and delivery robots, for example. But before these technologies become ubiquitous, we can make some deliberate choices about what this future accommodates.

Law and robotics scholar Kristen Thomasen has pointed out that robots raise urban design questions, including the commercialization or surveillance of public spaces—at the possible expense of some community members over others. In the United States, commercial food delivery robots that use public sidewalks have created obstacles for the people they share the space with, including obstructing the path of people who use wheelchairs. An organization in San Francisco came under fire in 2017 for using a 400-pound security robot to chase homeless people away from the space near its property. Governments have also begun to encroach on the idea of public space as an area of free assembly and expression by using drones and other surveillance technologies to dissuade protesters, while at the same time restricting or banning the use of private drones that can capture aerial imagery of police brutality.

Architectural choices for shared spaces are almost always made by people with power. (Many of our modern urban spaces are colonial constructs built on stolen land.) But if the public recognizes that these are political choices, we can drive them, too. Instead of letting commercial interests guide urban design, we could lobby to introduce robots into our cities with the goal of making urban areas more accessible to the public. Taking cues from pack animals and guide dogs, we could, for example, design autonomous vehicles and crosswalk robots, and the infrastructure around them, in ways that give people of different abilities more freedom to move around.

Generally, as we improve our autonomous systems, developing robots that have better sensors, processing power, and the ability to learn, it's clear that comparing them to humans is limiting. It prevents us from seeing a range of possibilities that we will encounter, and that we can work toward. As I hope this chapter shows, we can think differently about how to integrate this new breed, and a new analogy can open our minds to more possibilities for using robots in our workplaces, as well as to choices we have in our economic and political systems, and in the design of our physical world. But the power of challenging the human-robot comparison doesn't end here. It also influences whom we blame when things go wrong.

3

TRESPASSERS

ASSIGNING RESPONSIBILITY FOR
AUTONOMOUS DECISIONS

[Content warning for this chapter: death, animal cruelty]

"[T]he machine like the djinnee, which can learn and can make decisions on the basis of its learning, will in no way be obliged to make such decisions as we should have made, or will be acceptable to us. For the man who is not aware of this, to throw the problem of his responsibility on the machine, whether it can learn or not, is to cast his responsibility to the winds, and to find it coming back seated on the whirlwind."

—Norbert Wiener, mathematician and philosopher

"Where my beasts of their own wrong without my will and knowledge break into another's close, I shall be punished, for I am the trespasser with my beasts."

—Anonymous case during the reign of Henry VII

In the depths of the winter of 1457, in the town of Savigny-sur-Etang, in Bourgogne, France, a five-year-old boy named Jehan Martin was murdered. Witnesses reported the horrific crime to the authorities, and they were able to arrest the killer, who was a mother herself. Some of the witnesses suspected that her six children had participated in the crime, so the children were tried for murder alongside their mother at the local criminal court. After lengthy proceedings, the court found the accused murderess guilty, but there wasn't enough evidence from the eyewitness accounts to prove the children's involvement, and the court also decided that they were too young to punish. Their mother was sentenced to death by hanging.

It may seem unusual that the individuals on trial in this case were a sow and her piglets. But animal trials were a fairly common occurrence at the time.

Today, as machines that can make autonomous decisions enter our daily lives, many view the harm they cause as a new challenge to our legal systems and our ideas of responsibility. We tend to forget that this situation isn't brand new. Animals can act in unexpected ways of their own volition, causing harm with their decisions—just like robots.

We no longer put pigs on trial, because it's not rational to hold animals morally accountable for their actions, even if we've been seduced by the idea in the past. But our subconscious comparison of robots to humans, as ridiculous as it may seem, is nudging us toward the type of systems that treated pigs like people. The danger is that this allows companies, governments, and individuals to distance themselves from responsibility. It may seem preposterous today to hold a robot morally accountable for its actions, but there are already glimpses of this in how we talk about robot-caused harm—in ways that risk assigning them more agency than appropriate.

Fortunately, we have a long history of dealing with the question of animal-caused harm and have come up with a variety of other solutions, many of which are helpful to thinking through responsibility for harm in robotics.

WARNING: ERROR

In 2015, a robot technician named Wanda Holbrook was performing a routine repair at a bumper and trailer hitch factory in Grand Rapids, Michigan, when a robotic factory arm malfunctioned, fatally striking her in the head. Holbrook was in a section of the manufacturing floor that was separated into cells with safety doors designed to prevent the heavy machines from operating when a human was nearby. But the robot from the next cell didn't get the message and suddenly attempted to perform an operation in the cell she was standing in, something that her husband's 2017 lawsuit contended shouldn't have been possible. He sued a total of five companies involved in the design, build, and installation of the robot.

We are at the very beginning of the era of human-robot interaction. Robots, finally smart enough to navigate shared areas, but not smart enough to replace people, will be joining us in our homes, workplaces, and public areas. But factory robots cause accidents, robotic surgery has gone awry, and in 2016, a dystopian 340-pound security robot knocked over a sixteen-month-old in its path at a shopping mall (fortunately, the child suffered only minor injuries). Not only can these devices be physically dangerous; they are also driven by code, which can have some fairly consequential bugs, as evidenced by the 2018 and 2019 Boeing 737 Max plane crashes that were traced back to faulty software. The same is true for robotic cars, delivery drones, semiautonomous weapon systems, and household robots that are entering into shared spaces.

Because this is a book about physical robots, I'm going to focus on direct physical harm. It bears mentioning, however, that there are many other harms that can come from how we utilize algorithmic decision-making: search algorithms that reinforce racial or gender biases; the tin-eared response of virtual assistants to a mental or physical health crisis; artificial intelligence programs that issue unfair risk scores used in courtrooms; and hiring algorithms that put job candidates from certain demographics at a disadvantage. There's also overlap between this type of harm and the harm caused by physical robots, for example, if an autonomous weapon system were to use biased facial recognition. These important issues are outside the scope of this book, and, fortunately, others are calling attention to them: for a deep dive on harms in artificial intelligence and its integration, I highly recommend the work of researchers like Safiya Noble, Joy Buolamwini, Timnit Gebru, Ruha Benjamin, Latanya Sweeney, Sasha Costanza-Chock, Cathy O'Neil, and others, listed in the references to this chapter.

Machinery and product-related accidents in our world are not new (and neither are software failures, like the time the world nearly faced a Cold War disaster in 1983 when a Soviet system malfunctioned and

sounded the alarm that the United States had launched an intercontinental ballistic missile strike). The laws that govern liability differ across
countries; my European friends like to mock explicit warning labels on
US soda vending machines about risking personal injury if the dispens-
ing machines tip over, as well as the messages on US fast food coffee
cups that indicate that the coffee within is "very hot." On the whole,
however, there are reasonable legal solutions everywhere that address
responsibility for physical harm, from product liability for exploding
toasters to punishment for war crimes. So why does an internet search
with keywords like "robots," "AI," "responsibility," and "liability" return
hundreds of academic and media articles with titles and headlines like
"Who Is Responsible When Robots Kill?"

Some describe our current situation as a "historic moment" because
our robots, unlike our toasters, can make their own choices. The concern
is that these machines are different from devices we've dealt with previously because they are driven by artificial intelligence that lets them
make autonomous—and sometimes unpredictable—decisions. Uncrewed
aerial vehicles can function beyond the operator's line of sight, and self-
driving cars, well, drive themselves. Their manufacturers and users may
not be able to anticipate every consequence, if only because artificial
intelligence is able to plan and act independently, and even learn new
behaviors. Whose fault is it, and whose responsibility should it be, when
a robot malfunctions, or when it makes a decision that wasn't anticipated? Sometimes, it's not even possible for AI programmers to retrace
in retrospect how or why a certain decision was made.

Machines that can manipulate physical environments without
"understanding" when and how they're causing harm are scary to us.
Automakers are engrossed in massive research and development efforts
toward making autonomous cars commercially available, but in their
quest to put robot cars on the roads, their main challenge is public
concern. What types of decisions will the cars make, people ask, and
who is responsible for the harm caused by those decisions? Every year,
about 1.35 million people die in road wrecks. Fully autonomous cars
are predicted to be profoundly safer than human drivers, whose errors
cause more than 90 percent of car accidents. And yet, people don't like
the idea of letting the robots drive.

According to a 2018 survey by Allianz Global Assistance USA, only 43 percent of Americans are even interested in using self-driving cars, a number that actually decreased from 2017, which in turn decreased from 2016. (A separate Pew survey in 2017 confirmed that more than half of Americans were worried about autonomous cars, and 56 percent wouldn't even want to ride in a driverless car if given the chance to do so.) Daniel Durazo, director of communications for Allianz Global Assistance USA, put the numbers in context: "As our Future of Travel survey last year indicated, more travelers would feel safer on a rocket to space than being a passenger in a self-driving vehicle." In 2018, after a woman was struck by a self-driving car in Arizona, one of the few US states that has allowed testing on its streets, people started attacking the autonomous cars they encountered on the road, throwing rocks at them and slashing their tires.

On the way back from my visit to Victor Frankenstein's college town, Ingolstadt, I took my pregnant body to a cocktail party in Munich with a group of academic and industry folk in robotics, AI, and adjacent fields. As we mingled in an open-space loft filled with beautiful greenery, an older man in a suit struck up a conversation to tell me about his startup company. When he asked about my visit and I started to describe the autonomous vehicle accidents that car companies were grappling with, he interrupted me and recommended that I read the works of Isaac Asimov. When I told him that I was familiar, he narrowed his eyes in mock suspicion and asked, smiling, "But have you read *all* of his books on robots?" I started to respond that yes, I had read the robot series a few times, but he was too eager to make his point. He interrupted me again to describe the science-fictional concept well known to most people in robotics: Asimov's three laws for making robots safe for human interaction.

1. "A robot may not injure a human being or, through inaction, allow a human being to come to harm."
2. "A robot must obey the orders given [to] it by human beings except where such orders would conflict with the First Law."
3. "A robot must protect its own existence as long as such protection does not conflict with the First or Second Law."

This was not the first time that someone had recommended Asimov to me, and I felt a weariness descend, compounded by the fact that my body was still in a constant background process of 3D printing a tiny human being. That night, I was in no mood to listen politely or be talked over, so I excused myself and went to get a nonalcoholic cucumber blackberry spritzer. Don't get me wrong—I believe Asimov's work is iconic. When discussing his life and legacy, people often gloss over the fact that he was also well known for groping women at conferences and driving female sci fi writers out of the literary community, but he's probably *the* science fiction writer most associated with robotics. What I didn't have the energy to explain that night was that Asimov's stories are mostly about how programming robots with the "laws" doesn't work.

A surprising number of people, when confronting the question of harm in robotics, fixate on the idea of putting simple ethical rules into machines so that we can hold them accountable. (Asimov himself complicated things with a fourth law much later in his novels, but the Venn diagram between the people who know this and the people who know better than to apply the laws to actual robots tends to be more or less a circle.) As Maja Mataric, director of the University of Southern California Robotics and Autonomous Systems Center, has pointed out, Asimov's laws are "not something that [is] taken seriously enough to even be included in any robotics textbooks, which tells you something about [their] role in the field."

The nonlinearity of intelligence aside, people much smarter than me have spent decades and created entire fields of study to explore ethics in robotics. The subfield of machine ethics is especially dedicated to the idea of programming ethical rules into robots. Asimov was prescient: the idea of programming "ethical" decision-making into machines is a complex proposition. Whether we want robots to follow Asimov's laws, the Ten Commandments, or a model of utilitarian moral philosophy, translating those ethics into code has proven a seemingly impossible task.

One question that's gotten some attention is what decision an autonomous vehicle should make when faced with a trolley problem: the choice between running over five people by staying on a predetermined path or making a conscious decision to change direction and kill someone else in the process while saving the five.

The trolley problem, which exists in many different forms, is a philosophical thought experiment that's not designed to be solved, but rather to make us think about a tension in our moral choices—whether our actions should be based on inherent morality or on their consequences. Attempts to come up with functional rules—such as crowdsourcing the answer to whether a driverless car should run over a criminal to spare a doctor or letting the individual car owner decide what ethical rules their car should follow—have run into serious practical and philosophical hurdles. This debate also presupposes that it will be technically possible to effectively program whatever moral rules we humans agree on, provided we do ever agree. Human ethics are complicated. Even Wendell Wallach and Colin Allen, authors of *Moral Machines: Teaching Robots Right from Wrong* and proponents of the machine ethics approach, say, "The near future of moral machines is not and cannot be the attempt to recreate full moral agency."

If our laws are ill-equipped to deal with autonomous behavior, and we can't make robots follow moral rules, how should we handle this "historical moment?" I have my doubts about the idea that there is absolutely no precedent for this, and I worry that we are again viewing robots as some sort of humanlike agents that need to be accountable for their own decisions. It's strange to me that in robotics conversations around responsibility for physical harm, animals are remarkably absent. There are some differences between animals and robots, but an animal can plan, make its own decisions, act independently, and learn.

We've often felt like there's a difference between plain old bad luck or a random act of nature, like a lightning strike, and an instance like a pig attacking a child. At the same time, we don't expect animals to be able to abide by our laws and morals, the same way that we don't expect a toddler to reason like an adult. Like children, some animals are in a person or organization's realm of responsibility—someone who has the ability to reason and make informed choices (for example, a tiger may belong to a zoo).

Even though we don't hold animals morally accountable for their actions, we do acknowledge their ability to act on their own. Historically, as we've utilized animals for transportation, delivery, household help, and weaponry, we've often grappled with the question of responsibility for harm. When a circus company's elephant injures an abusive trainer, or a beekeeper's selectively bred honeybees join forces with a swarm of wild bees and attack the neighbor's dog, who is responsible? Animals can be owned by individuals or by organizations (or by no one), and some animal behavior results directly or indirectly from breeding or training.

It's a challenge to figure out how to assign the blame when something goes wrong, but it's one that's been around for millennia. When we look at our history of dealing with animal harm in Western law, we find a smorgasbord of creative answers. And, while most modern legal systems tend to treat animals more like property than people, they often acknowledge this "property's" penchant for autonomous behavior.

Today, as robots start to enter into shared spaces and knock over our toddlers, it is especially important to resist the idea that the robots themselves are responsible, rather than the people behind them. I'm not suggesting that we treat robots exactly like animals under the law. I am suggesting that there are more ways to think about the problem than trying to make the machines into moral agents. Trying on the animal analogy reveals that this is perhaps not as historic a moment as we thought, and the precedents in our rich history of assigning responsibility for unanticipated animal behavior could, at the very least, inspire us to think more creatively about responsibility in robotics.

THE CASE OF THE GORING OX

The Bible is packed with stories of oxen. They destroy crops, trample on people, and brawl with each other. A large body of historical and legal literature is concerned with the rules for what happens when an ox gores somebody, and for good reason. Oxen are among the first domesticated animals and were hugely meaningful to agrarian societies. Even today, cattle are among the most important animals we cultivate. So it's no surprise that oxen are the main example used to illustrate animal accountability throughout most of Western legal history and that the very origins of animal responsibility rules begin with the ancient case of the goring ox.

The Code of Eshnunna and the Code of Hammurabi, dating from

around the eighteenth century BCE, are some of the earliest extant "laws" known to humankind. We're not even sure we can call them laws because we don't know how they were enforced, but they are the first known rules created by Mesopotamian societies that lay out clear consequences for when an animal causes harm. Both "laws" established penalties for oxen owners, and both followed the same principle: if your ox kills someone unexpectedly, you're not responsible. But if you knew that the animal was a "habitual gorer"—in other words, that there was a risk—you're held to task.

Biblical law, created much later, also addresses the case of the goring ox. (It's worth noting that many historians believe that these rules were either copied from each other or originated from the same source.) Both the Mesopotamian codes and the Book of Exodus hold the owner accountable if, and only if, the owner knew that the ox had a tendency to gore people and failed to take proper care to prevent it.

These are the first known and recorded instances of assigning responsibility in the case of an autonomous agent. These rules attempt to take into account that oxen can act of their own accord. Because the human isn't expected to anticipate or control every autonomous decision made by the animals, the laws only fault owners who are careless with their beasts. In thinking about robots today, it's important to remember that responsibility for autonomous actions doesn't need to be an all-or-nothing question. From the very beginning, our laws have tried to strike an appropriate balance between fault and harm prevention.

Ancient Rome, the birthplace of Western law, had similar rules for goring oxen, but it also created a concept called noxal surrender: handing over the delinquent animal (or child, slave, or object) to compensate the person who was harmed. I'm not convinced that this concept would be useful if applied to robots today (in fact, it alarms me because I work with a robotics student who loves robots so much that it would probably incentivize her to throw herself in front of them, hoping to get injured). But it illustrates that we have found many creative solutions for how to deal with autonomous harm. There have also been other variants, like the "deodand" in England that had owners surrender their animals (and later their monetary value) to the state. Although these surrender laws aren't particularly well suited to modern robot

The Code of Hammurabi, which was found written on a giant stone, is one of the best-preserved and well-known insights into the rules from ancient Mesopotamia

cases, some of our ancient legal instruments for animals are worth considering today, for example, the idea that people need to take proper care and are only off the hook for harm if the robot's (oxen's) behavior was unlikely. We've also expanded on that basic concept, making our rules dependent on the context of our physical surroundings and even creating different legal categories of animals.

CATEGORIES AND FENCES

Animal law often distinguishes between situations where a human is only responsible if they could have anticipated the harm and failed to take measures to prevent it (like the laws of the goring ox) and ones where a human is "strictly" faulted for an animal's behavior. For example, a case during the reign of Henry VII establishes: "Where my beasts of their own wrong without my will and knowledge break into another's close, I shall be punished, for I am the trespasser with my beasts." The choice between these two options can depend on the context.

The Industrial Revolution was a time of utter chaos in England when the explosive growth of metropolitan manufacturing converted country dwellers into full-fledged city slickers, and they brought the animals along for the ride. Many people lived under the same roofs as their cows and chickens. Pigs wandered the streets, wreaking havoc as they brushed embers into straw and started fires or attacked and viciously bit other animals and children. The English needed to create laws for wandering livestock, and they needed to be rigorous. Owners had to fence in their animals, and they were held strictly responsible for harm if their livestock escaped, even if their fences were perfectly built.

Meanwhile, in the United States, where space was far more abundant, opposite norms started to evolve. There was a custom of letting animals graze over large swaths of land without fences. Some states refused to adopt English law and instead viewed it as the landowner's responsibility to build a fence if they didn't want others' livestock causing harm in their fields. Even as this norm was gradually eroded by legislation, responsibility was not strict. For example, when a horse escaped from a poorly constructed barnyard enclosure and struck a car on the road in 1926 in Wisconsin, a court decided that the owner couldn't have foreseen the animal's out-of-the-ordinary behavior and declined to hold him accountable.

With robots, our rules can be based on similar contextual considerations. For example, rules for robots that roam our streets and skies should be different in cities versus less populated areas, like the strict responsibility placed on animal owners in crowded English streets where the likelihood of harm was much higher.

Aside from the circumstances of our surroundings, another way we've taken the likelihood of harm (and how strict to be) into account is by creating different categories of animals. The ancient Romans believed that animals weren't capable of doing legal wrong themselves, but they differentiated between domestic and wild animals when creating rules for the owners. *Domestic* animal owners were only accountable if their furry culprit was "moved by some wildness contrary to the nature of its kind." The same was true in strict England, where the legal system made an exception for cats and dogs. Unless dog owners knew that their pets had "vicious propensities," they weren't necessarily held to task for damage caused by their trespassing canines.

We have similar rules in most places today. The idea is that wild animals are so inherently dangerous that they can't be made safe. This strict responsibility doesn't apply to pets that are generally predictable: if your fluffy little Pomeranian shows some surprising aggression and attacks a street performer on your walk, that's a different story than if you have a pet cheetah that mauls the poor, unsuspecting artist. The cheetah's behavior is going to be your "fault" pretty much regardless of whether you were careful with its leash, trimmed its claws every night, or trained it to be "safe."

It's certainly the case that some wild animals can be tamed, making them less suspicious of people. For example, the Bedouin have tamed gazelles to be comfortable enough with humans to be kept as pets. But this isn't always successful: certain animals have instincts that make them difficult to tame entirely. In 2003, magician Roy Horn (of the magician duo Siegfried & Roy) was attacked and severely injured by his pet tiger. So, "domesticated" generally refers to the tame (usually) vertebrates that depend on humans for survival—and "wild" creatures include any animal that isn't considered domesticated.

Categorizing animals by type, as "wild" and "domesticated" animals, helps create some nuance. Most animals are easily sorted into these two legal buckets, but there are always edge cases. For example, bees are somewhat predictable creatures and sometimes have a form of owner-ship attached to them, but they are mostly untamed, come and go freely, and can randomly sting people and animals, sometimes fatally. Because of their tricky nature, we've created special rules for bees for thousands of years, back to at least the Roman Empire. Many legal systems treat bees as fully or semidomesticated, despite their untamed nature. The main reasoning here seems to be that the bees aren't overly mischievous or dangerous creatures.

According to an English review of the laws on bees, "few animals are more prone than bees to furnish lawyers with attractive little problems." Courts have needed to decide whether bees can be owned at all and whether and when they can escape ownership. According to German law, "Where a swarm of bees takes flight, it becomes ownerless if the owner fails to pursue it without undue delay or if he gives up the pursuit." But even though most beekeepers officially own their bees, they aren't necessarily held responsible for their bees' autonomous roaming, trespassing, or even attacking. In 2010, the state of West Virginia (whose state insect is the bee) passed a law that explicitly protects beekeepers from responsibility for "ordinary negligence."

The edge cases show that it's not always easy to create categories, but that doesn't mean we haven't tried. And we could consider the same for robots, separating those that don't pose a lot of danger to people from more risky devices. New types of robots can emerge far more quickly than new kinds of animals, but the law is no stranger to cases that don't neatly fit into legal boxes, and it would be a useful exercise to try to categorize robots by capabilities (and risks). As robot ethicist Peter Asaro has pointed out, a robot vacuum cleaner may suck up and damage a piece of expensive jewelry, but it's not an inherently dangerous technology like an autonomous car.

With regard to bees, the Commonwealth of Virginia has gone so far as to put an official apiarist in charge of overseeing the state's bees, responsible for things like beekeeper education and ordering quarantine for specific bees. Creating a formal position of authority with specific expertise is one way to stay on top of harm prevention, and legal scholar Ryan Calo has proposed something in a similar vein for robots, calling for a dedicated federal robotics commission.

Rather than program morals into machines, our harm prevention could also follow some of the other methods we've used for animals— for example, taxing or regulating certain kinds of ownership or use.

DOG LICENSES AND SHEEP FUNDS

Every fall, the beautiful city of Vienna, Austria, spends two and a half weeks hosting a large festival called the Wiener Wiesn-Fest. The event bills itself as a celebration of Austrian tradition and folk music. Perhaps this falls under the umbrella of "tradition," but one of the main activities that the official festival descriptions fail to mention is plenty of drinking—in other words, people get plastered. During the 2018 Wiesn, in usual fashion, the police were called in to handle a brawl. An inebriated thirty-nine-year-old festivalgoer had gotten aggressive and also started to assault the police officers.

The officers responded with a strategic maneuver: keeping the police dog in a muzzle but ordering it to rush the man. But when the festivalgoer drunkenly grabbed at the dog's head, he ended up tearing off the muzzle, landing himself in the hospital with bite wounds. In this case, Austria declined to hold the police officers accountable for the incident. But if the dog had been in the control of another civilian rather than an officer of the law, the case might have taken a different turn. At the time, after a slew of dog-related injuries in Vienna, Austrian politicians were in the process of tightening their already comparatively strict regulations on privately owned dogs.

In Vienna, there's a specific list of dog breeds that are deemed dangerous, including pit bull terriers and rottweilers. Any controversy around the actual danger of these dogs aside, in making a distinction between "dangerous" and "non-dangerous" dogs, the government created stricter responsibility rules for certain harms, and also imposed a number of preventative measures to reduce risk, many of which are interesting to consider when we think about "dangerous" robots. Owners of "dangerous" dogs are required to be in control of their pups at all times, which includes being subject to the same strict alcohol limits as automobile drivers. Anyone in possession of a designated danger dog and tipsy enough to fail an alcohol test while out and about faces steep fines. The listed dogs can't be bred and can only be owned under stringent conditions, including using leashes and muzzles in public and having a special license that requires mandatory training sessions for the owners, who are tested again after two years. Anyone walking the dog also needs to have a license of their own. Rule breakers face hefty fines, and the dog is confiscated after their second violation. Dogs that aren't on the "dangerous" list but have previously bitten a person require licenses as well.

Not every city regulates dogs to this extent. Rules often differ depending on the country, the type of dog, and the individual community. In many European and Asian countries, specific dog breeds are considered dangerous and are banned or restricted. Requirements for ownership can include property signage to warn people that a dog is dangerous, as well as technological measures like special collars or even microchip implantation for identification. Dogs often need to be registered, and in the case of canines that are categorized as dangerous, some communities additionally require a formal notification if the dog is lost, passes away, or changes residency. There can be substantial fines if an illegal or restricted breed of dog causes harm, and often the dog will be confiscated or put down by the state.

In the same way that we've done for animals, we could consider what robot regulations could ameliorate some of the harm that creators or users can't anticipate. We could implement rules and design standards for robots in public and private spaces that have nothing to do with asking the robot itself to make moral decisions. Instead of only relying on their programming (i.e., "training"), we could put rules in place to ensure that their physical build is made safer (the robot equivalent of muzzles), that their handlers are able to control them (like the leashes and licenses mandated for dogs), and could consider a disclosure of risk (e.g., requiring signage on the property where the dog or robot is kept). Registration or microchip identification for specific classes of robots, mandatory pilot training for drones of a certain caliber, and restricting operation to permitted areas are just a few examples.

With animals, we've also created systems that help compensate for harm when it does occur, like taxes and insurance. While eighteenth-century England didn't hold dog owners strictly accountable, dogs were gradually becoming a public nuisance. After a particularly nasty outbreak of rabies, England reduced the widespread harm through another preventative measure: enacting a tax on dogs in 1796. Similarly, modern governments may implement registration fees or require insurance for dogs. For example, as farm communities grew in the US, some states created legal instruments to compensate farmers for the loss of their livestock to dogs, including "sheep funds." A sheep fund is a form of insurance: a pool of money collected from dog owners (usually through dog license or registration fees) that's distributed to anyone who loses a sheep as the result of a dog attack. Since attacks on sheep were fairly common, and it often wasn't possible to find the exact dog responsible, these funds compensated farmers without identifying the specific dog culprit and could be extended to compensate for harm by animals for which no one was responsible. Ohio, for example, created a system to give livestock owners a claim for compensation when an animal was injured or killed by a coyote.

By now, we've created many more types of "sheep funds" to insure against all sorts of animal-caused damage. Some countries impose insurance requirements on specific animals, such as mandatory expanded liability insurance for the owners of "dangerous" dogs. Insurance can also help in cases where multiple parties might be responsible. What happens when an animal trainer, rather than the owner or primary caretaker, trains an animal in a way that leads to problematic behavior? Trainers can teach a dolphin to locate and attack human divers. Dog owners hire professional handlers for dog shows; jockeys ride horses over the finish line. We have breeders, trainers, pet sitters, individual owners, corporate owners, and government owners.

Insurance doesn't necessarily redefine who gets blamed for bad animal behavior, but it does provide a mechanism to smooth over some of the complexity and compensate for financial losses, particularly as the insurance industry has developed internal norms for handling claims and only requires complicated legal responsibility analyses in edge cases. Like the sheep funds designed to simplify compensation for harm, we can create insurance systems that smooth out some complexities associated with compensating for harm to property, the way that we've already done for vehicles.

Humans have dealt with the question of animal harm from ancient oxen to Industrial Revolution pigs to modern pit bulls. We've implemented a number of rules and incentives that are designed to hold people accountable where appropriate and reduce overall harm. One thing is clear from this pile of legal rules: the problem of responsibility for autonomously caused harm is not new or unsolvable, and we could be drawing on our rich history to inspire appropriate rules for robotics. But, as I mentioned earlier, there's one solution we've chosen in the past that no one would deem appropriate for animals today, and it offers a possible answer to what is preventing us from thinking creatively about robot-caused harm: putting the animals themselves on trial.

THE ANIMAL TRIALS

Both the ancient Mesopotamian and Roman rules allowed for situations where nobody was responsible for an animal's unpredictable behavior, but for the Israelites, this idea didn't fly. Biblical law didn't necessarily fault the *owner* of a one-time offender, but the ox that fatally gored a person was stoned to death—an exceptionally harsh punishment that was usually only applied to serious crime cases. Unlike Mesopotamian society, which didn't create a stark separation between humans and the rest of nature, the Israelite system of law was based on religious beliefs that viewed humans as divine images of God placed above beasts. If an animal harmed a human being, even a "lesser" human being like a slave, it needed to be punished.

Other ancient laws contain similar implementations of retribution. The religious laws of the Vendidad from around the eighth century BCE prescribed punishment for mad dogs that wounded humans or sheep. The Mishnah, created between the second century BCE and the second century and often designated the first document of rabbinical Judaism, mandates that animals stand trial before twenty-three judges and, if found guilty, should be sentenced to death.

Some limited evidence exists that the Greeks also put animals on trial for murder, punishing them with death or exile. But the most remarkable and well-documented example of punishing animals for harm are the animal trials of the Middle Ages.

In no place is the strange history of animal trials better known than in the city of Basel, Switzerland. Even in our modern times, the city teems with monsters. Basilisks perch on fountains, glare at the world from paintings, and stand immortalized in statues on bridges. The scary, green half-bird, half-reptile is Basel's heraldic mascot and is featured everywhere. According to the lore, the basilisk was a feared beast that could kill people by breathing on them or instantly slay them with its evil glare. Basilisks allegedly hatched from eggs that were laid by roosters and were incubated by toads or snakes. So, in the year 1474, when a rooster (which, in retrospect, was probably mislabeled) committed the atrocity of laying an egg, the city of Basel prosecuted the fowl for the crime. In an elaborate formal trial, the rooster's lawyer (yes, the rooster had a human defense attorney) attempted to defend his client by argu-ing that it wasn't the rooster's fault this had happened: the act of laying an egg was unpremeditated and involuntary. But the court sentenced the rooster to death—shortly thereafter, the animal was burned at the stake in front of a large crowd of onlookers.

Today, despite the basilisk's popularity as a town mascot, it's unlikely that anyone in Basel believes that basilisks are real or that roosters should be taken to court, regardless of the heinous offenses they com-mit. It seems pretty far-fetched for anyone to want to legally prosecute a rooster or pig in our day and age. And yet we punished rogue beasts for hundreds of thousands of years of human history.

In the Middle Ages, animal trials swept through continental Europe, with most known cases in France, Germany, Switzerland, and Italy. Trials were also held to a lesser extent in Spain, Denmark, Turkey, and the Scandinavian countries, as well as in other parts of the world.

All sorts of beasts were put on the stand: goats, dogs, cows, horses, sheep, foxes, wolves, and more, but the most frequent criminal was the pig, probably because they were ubiquitous at the time and were often just left free to roam the streets, where they would attack small chil-dren. The earliest recorded pig trial was held in Fontenay-aux-Roses, near Paris, in 1266, where a pig was publicly burned for eating a child. Convicted pigs were buried alive in Saint-Quentin, hanged in Rou-magne, imprisoned in Pont-de-l'Arche, and tortured in Falaise. The French even hanged a pig in 1394 "for having sacrilegiously eaten a consecrated wafer."

Aside from putting individual animals on trial, which was gener-ally a secular proceeding, the church started performing another type of animal trial, wherein it intervened to deal with untamed animals that caused public harm—plagues of crop-destroying insects, for instance—bringing defendants ranging from mice and moles to cat-erpillars, worms, leeches, beetles, and snails to the holy stand. When weevils infested the vineyards of St. Julien, in France, the proceedings of this single case lasted for over forty-one years and filled twenty-nine pages of court records, including suggestions like giving the weevils their own plot of land to ravage in peace.

What's most striking about the animal trials is how meticulously

Trial of a Sow and Her Piglets

they followed the same legal process as for humans. An individual or a town could file a complaint and instigate a formal criminal investigation. Not all animal defendants were found guilty and punished—some were pronounced innocent or given reduced sentences. The animal trial process was taken so seriously that the towns even paid for a defense attorney for the accused creature(s).

But in the case of plagues, getting them to court wasn't easy. To let the invasive bugs or rodents know that they were in legal peril, ecclesiastical courts would send some poor court official to loudly read out the summons in public areas where the creatures were most likely to hear it. If the varmints didn't appear for their trial, they were pronounced guilty by default, but the court held session three times, sending out three separate warnings to the animal defendants, before issuing a default judgment.

When rats destroyed barley crops in a French town called Lucenay around 1508, the vermin were summoned to a trial and assigned a young lawyer named Bartholomy de Chassenée to defend their case. As usual, the summoned animals didn't show up for their court date. But Chassenée argued that not all the rodent defendants had heard the summons and convinced the court to reschedule the trial and instruct all the preachers in the surrounding area to issue summons for the rats at their local churches. When the rats failed to appear again, Chassenée, unfazed, held the line. He argued that it wasn't safe for the animals to attend their own trial, given the dangerous presence of cats, dogs, and unfriendly humans. Chassenée's bold arguments may well have swayed the court once again, but unfortunately for us, the final judgment and the fate of the rats are lost to history. Chassenée went on to defend many more creatures in court and published an authoritative book on insect injunctions. He was later played by Colin Firth in a 1993 British/French film called *The Hour of the Pig*. (A different version was released in the US as *The Advocate*, and the original was harder for me to get my hands on than the actual animal trial accounts.)

The practice of animal trials lasted for hundreds of years. Even after the formal trials petered out, capital punishment for animals remained an occasional event. In 1916, in Erwin, Tennessee, Mary the circus elephant attacked and fatally wounded an untrained handler who tugged at her sensitive ear with a bullhook as she was trying to eat a watermelon rind by the side of the road. She was hanged for murder, and the details are too horrifying for me to recount here (feel free to look it up if you have the stomach). According to some accounts, the Commonwealth of Kentucky sentenced a German shepherd to death for attempted murder and executed the dog via electric chair as late as 1926.

One of the benefits of animal trials isn't just that they satisfy a religious motivation or desire for retribution. When we treat the animals like humans, assign them moral responsibility, and have them stand trial for their own actions, we conveniently sidestep the puzzle of assigning indirect blame for autonomously caused harm. But, looking back, it seems ridiculous to most of us now to hold the animals themselves accountable for the harms they cause. We know that animals can't follow our moral rules, so it doesn't make any sense to blame them for their "crimes." And yet, we're seeing the same tendency toward direct blame today—in our conversations around responsibility for robot-caused harm.

THE NEW BLAME GAME

When Wanda Holbrook was crushed by the factory robot, the headlines read "A Rogue Robot Is Blamed for the Gruesome Death of a Human Colleague," "What Happens When Robots Kill?" and "Robot Goes Rogue and Kills Woman." Many online articles forewent photographs of the actual machinery, a standard factory manipulator, opting instead for images of metal humanoid robots to illustrate their "killer robot" piece. We know that the machine didn't intentionally cause the fatal accident that day, and yet we've somehow reverted back to a society that blames the pigs.

For decades, science fiction has raised questions about chains of causality and how to assign responsibility for harm caused by robots. And while I'm grateful to Asimov and others for kicking off that conversation, these pop culture references have inspired a very narrow range of proposed solutions. Vast swaths of media articles and academic work claim that our current liability laws for product-caused harm aren't enough and that we need to program human ethics and moral decision-making into robots to prevent harm. Some experts also propose that we should hold robots themselves accountable when harm does occur.

A 2016 draft report by the European Parliament's Committee on Legal Affairs suggested that the EU consider creating a special legal status ("electronic personhood") for robots that can learn, adapt, and act for themselves, making them liable for any harm they cause. The (ultimately rejected) report created a stir in the press and some scorn within the robotics and AI community, but it was by no means a new idea. Many an author, legal scholar, or philosopher has argued that future robots should be held accountable for harm. Today, they say, any harm caused by a machine can be directly traced back to the human or firm who designed, built, or programmed it, but in the future, robots will be capable of acting independently, and in ways that their creators can't anticipate.

Before completely dismissing the idea of holding machines accountable, remember that there is some legal precedent beyond pig trials: we've given corporations both rights and responsibilities. We already have models for creating nonhuman "personhood." But what worries me most about the robot personhood approach to the topic is that it seems mostly driven by science fiction, and in particular the robot-human comparison. Our desire to anthropomorphize—to project human feelings and behaviors onto robots, which I'll talk about in the next part of this book—isn't going away. At the same time, slapping a picture of the Terminator on articles about robot-caused harm and discussing how to create responsibilities for the robots themselves is limiting.

The fact that we immediately leap to the idea of holding robots morally responsible illustrates the depth of our quasi-human assumption. We too often forget the other ways we've addressed autonomous nonhuman behavior in the past. In 2018, the US Chamber Institute for Legal Reform published a report on addressing the liability and regulatory implications of emerging technologies. In it, they suggest alternate models of assigning responsibility as robots and AI become able to make their own decisions. In only a few small paragraphs of the ninety-three-page report, they suggest that one of these alternatives could be to treat robots like pets, an idea that's also popped up in a small slice of academic work, unfortunately largely overshadowed.

From oxen to pit bulls, our laws have long dealt with the risky behavior of animals, and, in some ways, robots are easier, since they're intentionally created by people and are always in somebody's purview. (Thankfully, we don't have feral coyote robots roaming the plains and killing our sheep, yet.) Why would we create a personhood for robots instead of thinking through our other options, from sheep funds to animal category and registration? Contrary to the popular narrative that we're on entirely new ground, our history with animals shows that we've deeply grappled with the question of autonomous agents that cause unanticipated harm based on their own decision-making.

At first blush, the idea of giving robots their own legal personhood seems like a simple way to sidestep the complexity of dividing responsibility among the people who build, program, train, and own robots, but it's not necessary. We've dealt with some divided responsibility with animals, like the fact that animals can be owned by a person, corporation, or government and that they can be trained by one person and handled by another.

We also have systems to figure out complicated responsibility situations among people, like in agent or employer-employee relationships. But while the legal principles we've developed for people are a useful precedent for us here, comparing robots to animals allows us to break out of the robot-human comparison mold. Most importantly, when we view robots through the lens of animals, the idea of assigning them human responsibility *feels* very different: we don't train dogs to understand and evaluate the moral complexity of war situations; we train them to attack tanks. And when they then attack the wrong tanks, we blame the trainers, not the dogs.

The animal comparison isn't perfect. Right now, autonomous cars are programmed to avoid all obstacles and are unable to "decide" to save a pregnant person, but robots might still be capable of different complex analysis and decision-making than animals. (Our guided missiles are better at seeking out specific targets than a flaming pig or bat bomb.) But the animal analogy helps us start from the baseline that robots don't make decisions like humans and that the responsibility for the harm they cause needs to be assigned to people. It lets us come at the problem from the assumption that robots are in our purview, *even if* they can make decisions we didn't anticipate.

Autonomous vehicles are unlike both horse-drawn carriages and modern cars in that the driver often won't be the main responsible party: in the case of robots, we'll often look to the companies who make and sell them. Like in the Wanda Holbrook case, multiple companies may be involved in developing risky technology, from hardware to software. For some robots, product liability law is an existing piece of the puzzle. Product liability holds anyone who makes products available to the public responsible for injuries that faulty products cause. For example, manufacturers and distributors in the United States have a strict obligation to disclose nonobvious risks—don't mix two cleaning solvents together, for example—and avoid any manufacturing mistakes. For robots, we may find ourselves confronted with cases where a robotic platform uses open source software with many anonymous contributors or where such systems are discouraged solely for liability reasons. But in these cases, we need to be weighing the pros and cons without putting agency on the robots themselves.

Some have argued that, like the animal trials, we should "punish" future robots for misbehavior, either because the robots themselves have developed the ability to suffer or because it satisfies people's desire for justice and retribution. As our history of punishing animals tells us, it's not unthinkable for society to want to punish a nonhuman. And our criminal law is far from an emotionless equation. But I have some additional concerns about the idea of holding robots themselves accountable for the harms they cause.

Our constant subconscious comparison of robots to humans can lead to what legal scholar Neil Richards and roboticist Bill Smart term the Android Fallacy, where courts and lawmakers could start to use humans as their analogy for robots in technology regulation. They point to the history of privacy law in the United States and how courts had to decide whether email was more like a postcard or a letter. The chosen analogy (letter) had profound implications for whether emails were considered private communication, changing the nature of our correspondence as a whole. While privacy scholars breathed a sigh of relief that courts chose the letter in this case, Richards and Smart use it to illustrate how much our analogies matter. Both they and legal scholar Ryan Calo have cautioned against our automatic comparison of robots to people, warning that it could lead to inappropriate legal doctrines.

I worry that by focusing on the robots and their unanticipated behavior, we will start to neglect the accountability of those who create them. According to Peter Asaro, one of the overarching goals of robot ethics is to prevent robots from doing harm, but also to prevent people from escaping responsibility for their actions. Our subconscious human comparison lends itself too much to the idea that companies and individuals shouldn't be held responsible for algorithmically driven behavior they didn't predict. And while it will make sense to let companies off the hook in some cases, "the robot did it" shouldn't be allowed as a new type of excuse. Even if we believe that, morally speaking, it's nobody's "fault" when a machine misbehaves, my background in law and economics screams, "Incentives!" at every legal decision point. Companies need to continue to be held responsible for the technology they put out into the world, even when the chains of causality are complex. We're already seeing people try to blame the robot with AI systems in play today, where corporations keep their algorithms as secret as possible, claiming protection of proprietary secret sauce while pointing fingers at data providers, users, and the system itself, trying to distance themselves from responsibility for unanticipated outcomes.

Companies will argue that by placing the risk of robot errors on them, we will stifle innovation. When Microsoft released a Twitter bot named Tay whose online learning was co-opted by web community 4chan and thus trained to spew racist hate speech, was the responsibility all Microsoft's? How will companies or individuals be able to experiment and innovate? While we do need to take care not to throttle innovation more than needed, every legal system in the world has the capacity for nuance, even in cases of strict liability. Speech can harm, and if Tay had physical abilities and could be trained to set fire to buildings, we would want companies to think twice about putting it out in the world and to think more deeply in advance about how it could be manipulated by people to cause harm. Most legal systems distinguish between setting a hamster loose versus setting a tiger loose, and we could do that here, too. A soft robotic pillow with a wagging tail (this is an actual product) is very different from an autonomous underwater vehicle that can shoot a gun (sadly, also an actual product).

Complicating this blame game, we're not going to see widespread situations where a human task falls to a robot (in which case the responsibility mostly just switches from the human to the robot manufacturer). Rather, we're going to see people working *with* robotic technology, and some of the harms caused may be due to shared mistakes. Even in the case of autonomous cars, we're looking at many years of shared-responsibility situations before we get anywhere close to the less complicated case of fully autonomous vehicles. In the situations we're facing, we need to be particularly aware of whom we hold responsible and why.

HUMAN IN THE LOOP

In theory, human-machine teams work well. Self-driving tractors can plow fields, so long as a farmer is on premises to deal with the unexpected hiccups. But as we automate our vehicles, it's not as simple as just letting a robot take the wheel. It requires rethinking a system and using human input where it's most valuable. Like horse-drawn carriages, if we can let a human use their broad intelligence to deal with unexpected situations, while letting the robot automate the predictable work, this type of team will function better than either human or robot alone. But as we've already seen with some of the automated driver assistance systems in modern cars, there's a big problem with figuring out the handoffs.

Multiple Teslas crashed over the past few years when their drivers weren't paying attention, smashing into trucks, concrete barriers, or, most recently, the back of a police car. Tesla warns its customers that the autopilot system doesn't replace the driver, but with a machine doing all the steering, braking, and accelerating, it's a lot to ask of a person to sit there and pay close attention to everything. This is still a problem in planes. Commercial pilots these days mostly sit around and monitor the status of the aircraft rather than flying the plane themselves. But when the system encounters something unexpected and fails, it hands off control of the aircraft to the human pilots. This has led to plane crashes that were the result of both machine and human error. And here's where the blame game gets tricky—in shared responsibility situations, the media tends to attribute the plane crashes to the human pilots.

The problem here isn't that our legal systems don't have a solution for this sort of liability—they do. The problem is that our *perception* of fault in these situations is often different. Madeleine Clare Elish calls this the "moral crumple zone," where human-robot interactions can cause outcomes that tend to deflect blame away from the machine (and thus its creators), even if the human operator isn't 100 percent at fault. We need to be extremely careful that this bias doesn't let companies deflect legal liability.

Our biases in assigning blame for harm are just one example of how we should be thinking about robots differently. In general, we need to maintain a clear-eyed view of the limitations and technological determinism that come from our conception of robots as quasi-humans and find ways to counteract our current default narratives. Using an animal analogy helps us to put our current systems, whether technological, economic, legal, or cultural, into perspective, and it opens our minds to alternate possibilities—something that is especially important as we continue to integrate robots into our world.

But the usefulness of this animal comparison doesn't end with workplace integration and responsibility for harm. The next part of this book looks at another area that is rife with science-fictional bias and moral panic: the near-term future of robot companionship.

II

COMPANIONSHIP

. . . . .

4

ROBOTS VERSUS TOASTERS

"Man is by nature a social animal."

—Aristotle

Around 2014, Hiroshi Funabashi received an odd request from a client. Funabashi was the supervisor of repairs for A Fun Co., a Japanese company that specialized in fixing old Sony products that had been discontinued or fallen out of warranty. A seventy-five-year-old woman had submitted a robotic dog called an AIBO for repair, but instead of describing a technical problem, she wrote in asking whether anything could be done for her robot's "aching joints." In his correspondence with the woman, Funabashi realized that she viewed the robot dog as her pet.

This wasn't accidental on Sony's part. The little metallic cyber-dog AIBO (Japanese for "pal" or "partner") was marketed as "Man's Best Friend for the 21st Century." Launched in 1999, AIBO was one of the most sophisticated interactive toys to hit the market. The little companion dog had a persona, reacted to people, and could act happy and sad. According to Sony's description, AIBO had "real emotions and instincts." That was, technically speaking, false advertising, but it didn't matter. People knew that their AIBO couldn't feel, and yet many of them treated it as though it did.

When Sony announced the first few limited runs of the toy in Japan and in the US, every unit sold out within minutes. Sony ramped up their production, releasing a variety of "breeds" in different colors over the next six years and selling 150,000 units worldwide. But in 2006, struggling to pare losses in the electronics business, Sony announced that they would be discontinuing AIBO. Eight years later, in 2014, they halted customer support for the AIBOs that were out in the world. The news devastated the remaining community of AIBO owners. To them, the news that Sony was pulling its repair services meant that their AIBOs would die.

Some of the AIBO owners, like the woman worried about her robot's "aching joints," discovered A-Fun Co. The Sony product repair company started to see an influx of broken AIBOs. They found themselves devoting time to re-create the AIBO's schematics in order to better service the robots. Soon, they were getting so many requests that they began to harvest parts of unfixable AIBOs and collect them to use as part replacements for the fixable ones. Nobuyuki Norimatsu, the former Sony engineer who founded the independent repair company, became sensitive to the emotional connections that people had to the robot dogs. A-Fun Co. started calling the AIBO part swaps "organ donations" to better capture how the owners felt about what was happening. He also started to think about how to honor people's loss of the unfixable "donor" AIBOs. One day, he reached out to Kōfuku-ji, a Buddhist temple, with an idea. The temple agreed to perform a ceremony for the robots that could no longer be fixed.

Kōfuku-ji held its first AIBO funeral in 2015. Seventeen robot dogs were thanked for their "organ donations" and bid farewell in a formal Buddhist ceremony, with prayers and chants. This would not be the only robot funeral at the temple—more AIBO owners wanted the opportunity to say goodbye to their companions. In 2018, the same year that Sony launched a new, modernized version of AIBO, Kōfuku-ji honored its 800th unfixable robot. A-Fun Co.'s founder, Nobuyuki Norimatsu, went on to explore a new venture called Robot Therapy, designed to help people cope with the loss of robot companions.

Funerals for robot dogs may seem par for the course for the Japanese. After all, Japan has a long history of Shintoism, an Indigenous belief that everything has a spirit: people, animals, trees, rocks, man-made artifacts, and even nothingness. Unlike Christian or Islamic societies that create a strong divide between alive and not alive, the Japanese have long-standing traditions of honoring objects, for example, holding funeral ceremonies for their possessions, from futons to broken sewing needles. It seems like a fairly logical extension to perform traditional mortuary rites for modern devices. But it wasn't just the Japanese AIBO owners who mourned the loss of their dogs. Even in the United States, with very different religious underpinnings, people felt connected to their AIBOs.

Genie Boutchia, a thirty-six-year-old stay-at-home parent, observed her children treat AIBO like a family member, but it was more than just the kids. "I always thought I was pretty rational but I don't think of her as a toy anymore," she said of their robot dog. "She's like part of the family. . . . It's so strange. You become attached. I know she's a hunk of plastic but she's just fabulous. . . . I really can't put it into words why I like her." Boutchia says she felt similarly toward the robot as she had felt toward the living dogs she had owned in the past. "One day, when I couldn't get her to boot up, I was sick to my stomach. I thought my head was going to explode." Grace Walkus, a publishing executive from New Jersey, describes a similar experience: "You talk to AIBO like you would to a real pet. You pet AIBO like you would a real pet and from there other emotions evolve. You feel guilty if you leave AIBO alone and it cries, and you think twice before scolding him or her." But AIBO wasn't alive, didn't have feelings, and couldn't understand human emotion. And Boutchia and Walkus knew this.

Before I explore what robot companionship means for us, this chapter delves a little deeper into the weird world that I've been immersed in for the past decade: why we treat robots like living things.

ANTHROPOMORPHISM (PROJECTING OURSELVES)

I'm not the only robot nerd out there. People are fascinated by the robot characters in our media and pop culture and the science fiction stories that imbue them with humanlike behaviors, desires, and personalities. Media, of course, wields a lot of influence over our perception. Many a child has excitedly shouted, "Curious George!" when they first see a Barbary macaque monkey at the zoo. Just like our toddlers, we tend to imagine characters based on the stories we're familiar with. Our rich collection of robotic personas in science fiction is evidently one reason we personify the robots in our lives: Rosie is one of the top names that people choose for their Roomba vacuum cleaner. So is science fiction to blame for our robot dog funerals? Not entirely.

The portrayal of robots in media is a chicken or egg question: do we personify robots because they're shown to us as characters, or are they shown as characters because that's how we perceive them in the first place? It may be the latter: our fascination with machines as agents goes back much further in our history than *The Jetsons*. According to historian Adrienne Mayor, author of *Gods and Robots*, we've been drawn to the idea of autonomous "robots" for millennia. So our reaction to robots seems grounded in something beyond just having grown up watching robot characters on TV.

Another guess as to why we treat robots differently than, say, toasters is that we lack the basic knowledge of how the technology works. As science fiction writer Arthur C. Clarke famously wrote, "Any sufficiently advanced technology is indistinguishable from magic." Inexperience with robots and their inner workings may make certain machine behavior seem magically lifelike to people. And in our current world, as robots move out from behind factory walls into shared spaces, our inexperience could be what's causing us to project our science-fictional notions onto our robotic devices.

This is what a scientist from a different field assured me. I was at a robotics and space conference in Palm Springs, California, getting a spiked lemonade at the bar outside, when I met Scott. The weather was gorgeous, the bar was empty, and we easily struck up a conversation as the only two people skipping out for a drink at two o'clock in the afternoon. Scott's area of expertise was the giant planets and the origin of the solar system. He had spent the past thirty years as a principal investigator at NASA, sending rockets on missions to other places in our universe. When we got around to the topic of human-robot interaction, he was interested. But when I mentioned that I think people treat robots more like living things than devices, his response was "Eh . . . I don't buy it."

Playing with his straw, Scott suggested that this was, at most, a generational issue. Maybe we do this now, he ruminated, but the next batch of people, the kids who grow up with robots all around them, will treat their robots just like any other device. We stayed at our table under a colorfully striped sun umbrella in intense conversation for two hours, barely noticing the balmy weather or our empty glasses of lemonade.

It's true that our generation names our robots R2-D2 because we love *Star Wars* and that some of us will reflexively blurt out, "Excuse me," when we first encounter a delivery robot on the sidewalk. As we start to interact with more robots in our workplaces, households, and public areas, our culture will surely change over time. And it's possible that people's tendency to treat robots like they're alive will fade away as our familiarity with the technology changes. But as I told Scott on that sunny afternoon—I don't buy it. Perceiving robots as living things goes deeper than a simple novelty effect, and it ties directly into the biological hardwiring that motivates us to see ourselves in others.

Why is Curious George a monkey? The character is written for children, but even adults constantly assign human traits to animals and objects. We have an inherent tendency to anthropomorphize—to project our own behaviors, experiences, and emotions onto other entities. We view monkeys as curious, sloths as lazy, and our cats as secretly plotting to overthrow us. We'll name cars, stuffed animals, statues, and sourdough starters. We see faces in inanimate objects, like eyes in buttons or car headlights, and we get annoyed at our "misbehaving" office printers. In 1975, American advertising executive Gary Dahl became a multimillionaire by selling pet rocks. The pet rocks' popularity was short-lived, but these fancies that we engage in with animals and objects persist, and they may even have a purpose.

It's not entirely clear when humans developed the ability to see ourselves in others. One guess is that it happened as our culture and relationships with animals and nature changed in the Middle to Upper Paleolithic transition, some 40,000 years ago. As for why we needed to learn to project our human traits, evolutionary theories range from needing to identify danger, to needing to form social alliances to survive, to scientists just assuming there must be *some* sort of reason for it, given that it has such strong manifestations and triggers.

Our tendency to anthropomorphize animals and objects is a stable characteristic that is robust over time and already present in young

children. One of the first things an infant learns to identify and focus
on is a face, whether it's their caregiver's or even just a black-and-white
drawing of facial features. Babies also seem able to create simulations
of other people's mental states from birth onward, an ability that may
contribute to how we, from childhood, perceive animals and certain
objects as social subjects. Evolutionary psychologists think one reason
we anthropomorphize is that it helps us understand the world around
us. We have a desire to reduce uncertainty and to be able to predict
what might happen in our social environment. This causes us to constantly try to make sense of the people, animals, objects, and concepts
we encounter. And in order to interpret these confusing others, we map
them after what we know best: ourselves.

Another popular theory behind our anthropomorphism is that it
feeds our inherent need for social connection. Screenwriter William
Broyles consulted with professionals as he was creating the character
Chuck, played by the Oscar-nominated Tom Hanks in the 2000 movie
Cast Away. Stranded alone on an island and starving for social connection, Chuck develops a personal relationship with a volleyball, naming
it Wilson and talking to it like a friend. His emotional connection to
Wilson is so strong that he feels deeply responsible for the inanimate
object, even apologizing to Wilson at the end of the movie.

Cast Away is fiction, but loneliness can be connected to anthropomorphism. People who are lonely appear to have a much stronger
tendency to anthropomorphize nonhumans, even bonding with objects
to the point of developing deep personal relationships. Social isolation
may also be one of the triggers for developing objectophilia, a romantic
or sexual attraction to objects. Erika Eiffel caused a media stir when she
changed her last name after "marrying" the Eiffel tower in 2007, and in
2016, Chris Sevier sued a county clerk for refusing to issue a license for
him to marry his computer. Others have developed emotional relationships to objects from amusement park rides to bridges to hi-fi systems.

But anthropomorphism goes beyond just loneliness and survival.
There are many far less extreme examples of behavior toward objects that
are similar to how we might treat other people. People will name their
cars, kiss dice for good luck, and apologize to their childhood teddy bear
before putting it away. We may sometimes feel silly for this behavior, but

we also seem to enjoy it. As Aristotle once said, we are by nature a social animal.

In the modern world of virtual characters, our emotional relationships to nonliving things have sometimes gotten serious. One of the biggest children's toy fads in the 1990s was the Tamagotchi, a virtual pet game that kids could carry around on a key chain. The simple character went through different life stages and required players to engage in a variety of "care" in order to keep it happy and healthy. If neglected for too long, the Tamagotchi would "die." The game became so popular and was so distracting that schools started banning it. Kids were treating the tiny character on the screen like something alive that needed to be nurtured, responding to its "needs" whenever they could. A teenage girl in Japan reportedly committed suicide over her Tamagotchi's passing. And it wasn't just kids. Parents sheepishly admitted to taking their children's Tamagotchis to work so that they could care for them.

Video game characters can inspire a similar level of loyalty, even when they are extremely simple in design. The popular 2007 video game *Portal* featured a virtual object called the Weighted Companion Cube, a simple 3D box with hearts on the side that accompanied the player through the game. In the final level, which required sacrificing the cube, some players chose to forfeit their victory instead of their companion.

When personal computing exploded in the 1980s, researchers became interested in people's behavior around these new, widespread electronic goods, and tech companies were invested in figuring out how to create efficient and positive interactions for the users of their products. It spawned a field of research on how people communicate with computers called human-computer interaction, or HCI. The researchers in this field discovered that it doesn't even take a constructed character to make people anthropomorphize computers.

In the late 1990s, some Stanford researchers asked experiment participants to complete a learning exercise on a computer. After the task, they asked them to rate the computer's performance. For this, they split the participants into three groups: one group gave their performance rating on the same computer with which they had done the task; one group rated the performance on a different computer; and one group did the rating on paper. The group assigned to the *same* computer rated the computer's performance much more positively than the other two groups. The only explanation the researchers could find was that people didn't want to hurt the computer's feelings. Their results indicated that the participants perceived the computer as a social actor and automatically applied the same rules that they would in any other social interaction: they were polite.

This study was one of hundreds that Stanford professor Clifford Nass, together with his colleagues and students, conducted on how people treat computers like social actors. Nass and his colleague Byron Reeves had begun to take traditional human-to-human psychology studies and replace one of the parties with a computer. They looked at behaviors like praise, cooperation, and reciprocity, seeing over and over again that our behaviors toward computers went in the same direction as our behaviors toward humans would. For example, if a computer helped with a task, people would reciprocate and devote more time and effort to helping it with something. Nass and Reeves coined the term "the media equation" to describe how people tend to treat computers like people and will mindlessly apply social conventions to their interactions with them.

The research in HCI also suggests where some artificial agents go astray. Microsoft's former Office assistant Clippy is possibly one of the most hated animated characters of all time. The cartoon paper clip with the waggly eyebrows was meant to be useful: if its software detected that a user was beginning a document with the word "Dear," Clippy would pop up and offer to help write a letter. But the hostility that people felt toward the "helpful" paper clip on their screens resulted in many complaints and inspired years of derogatory viral memes, Clippy paraphernalia, and even a (possibly parodic) book of the cartoon paper clip engaging in lewd acts. Nass's research helped Microsoft figure out what went wrong: users hated Clippy because they perceived Clippy as a social agent instead of a tool, and they expected the paper clip to follow social conventions. Instead, its social behavior was terrible. It would constantly bug people, not understand them, and come across as a spying nag rather than a helpful tool.

With the current rise of physical robots that interact with people, the findings in HCI have inspired a new field of research: human-robot interaction, or HRI. So far, a lot of the research in this new field shows the same trends for robots: people will mindlessly apply social conventions when interacting with them. From disliking robots that aren't "nice" to being more willing to collaborate with them when they are, and even feeling sorry for them when someone else is cruel to them, the same patterns are clear, and they're even stronger.

On the final evening of the robotics and space conference where I met my lemonade friend, Scott, we bumped into each other again while exiting a meeting hall. As we walked out into the starry evening, Scott said, "I'm starting to wonder if maybe you were right." He told me that he had stumbled across a robotics demo at the conference: a flying robot that was modeled after a dragonfly. As it buzzed and swooped over people's heads, he had looked around and noticed something interesting: everyone watching the demo was completely captivated. All eyes in the room were following the robot with rapt fascination as it moved through the air. "Maybe it's not a generational thing," Scott said to me, still not entirely convinced, but wavering; "maybe we're built this way." As we said goodbye, we shook hands and agreed to follow up on the answer in a few decades.

In the meantime, our response to robots, like the dragonfly demo, tends to be more extreme and more visceral than any of the social behavior we've witnessed in human-computer interaction. This is because physical robots trigger another piece of our biological hardwiring: our perception of movement.

THE POWER OF MOVEMENT

When the first black-and-white motion pictures came to the screen, an 1896 film showing in a Paris cinema is said to have caused a stampede: the first-time moviegoers, watching a giant train barrel toward them, jumped out of their seats and ran away from the screen in panic. According to film scholar Martin Loiperdinger, this story is no more than an urban legend. But this new media format, "moving pictures," proved to be both immersive and compelling, and was here to stay. Thanks to a baked-in ability to interpret motion, we're fascinated even by very simple animation because it tells stories we intuitively understand.

In a seminal study from the 1940s, psychologists Fritz Heider and Marianne Simmel showed participants a black-and-white movie of simple, geometrical shapes moving around on a screen. When instructed to describe what they were seeing, nearly every single one of their participants interpreted the shapes to be moving around with agency and purpose. They described the behavior of the triangles and circle the way we describe people's behavior, by assuming intent and motives. Many of them went so far as to create a complex narrative around the moving shapes. According to one participant: "A man has planned to meet a girl and the girl comes along with another man. . . . The girl gets worried and races from one corner to the other in the far part of the room. . . . The girl gets out of the room in a sudden dash just as man number two gets the door open. The two chase around the outside of the room together, followed by man number one. But they finally elude him and get away. The first man goes back and tries to open his door, but he is so blinded by rage and frustration that he cannot open it."

What brought the shapes to life for Heider and Simmel's participants was solely their movement. We can interpret certain movement in other entities as "worried," "frustrated," or "blinded by rage," even when the "other" is a simple black triangle moving across a white background. A number of studies document how much information we can extract from very basic cues, getting us to assign emotions and gender identity to things as simple as moving points of light. And while we might not run away from a train on a screen, we're still able to *interpret* the movement and may even get a little thrill from watching the train in a more modern 3D screening. (There are certainly some embarrassing videos

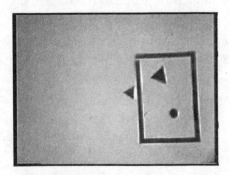

Screenshot from Heider and Simmel's 1944 animation

of people—maybe even of me—playing games wearing virtual reality headsets.)

Many scientists believe that autonomous movement activates our "life detector." Because we've evolved needing to quickly identify natural predators, our brains are on constant lookout for moving agents. In fact, our perception is so attuned to movement that we separate things into objects and agents, even if we're looking at a still image. Evolutionary psychologists Joshua New, Leda Cosmides, and John Tooby showed people photos of a variety of scenes, like a nature landscape, a city scene, or an office desk. Then, they switched in an identical image with one addition; for example, a bird, a coffee mug, an elephant, a silo, or a vehicle. They measured how quickly the participants could identify the new appearance. People were substantially quicker and more accurate at detecting the animals compared to all the other categories, including larger objects and vehicles.

The researchers also found evidence that animal detection activated an entirely different region of people's brains. Research like this suggests that a specific part of our brain is constantly monitoring for lifelike animal movement. This study in particular also suggests that our ability to separate animals and objects is more likely to be driven by deep ancestral priorities than our own life experiences. Even though we have been living with cars for our whole lives, and they are now more dangerous to us than bears or tigers, we're still much quicker to detect the presence of an animal.

The biological hardwiring that detects and interprets life in autonomous agent movement is even stronger when the movement has a body and is in the room with us. John Harris and Ehud Sharlin at the University of Calgary tested this projection with a moving stick. They took a long piece of wood, about the size of a twirler's baton, and attached one end to a base with motors and eight degrees of freedom. This allowed the researchers to control the stick remotely and wave it around: fast, slow, doing figure eights, etc. They asked the experiment participants to spend some time alone in a room with the moving stick. Then, they had the participants describe their experience.

Only two of the thirty participants described the stick's movement in technical terms. The others told the researchers that the stick was

bowing or otherwise greeting them, claimed it was aggressive and trying to attack them, described it as pensive, "hiding something," or even
"purring happily." At least ten people said the stick was "dancing." One
woman told the stick to stop pointing at her.

If people can imbue a moving stick with agency, what happens when
they meet R2-D2? Given our social tendencies and ingrained responses
to lifelike movement in our physical space, it's fairly unsurprising that
people perceive robots as being alive. Robots are physical objects in our
space that often move in a way that seems (to our lizard brains) to have
agency. A lot of the time, we don't perceive robots as objects—to us,
they are agents. And, while we may enjoy the concept of pet rocks, we
love to anthropomorphize agent behavior even more.

We already have a slew of interesting research in this area. For example, people think a robot that's present in a room with them is more
enjoyable than the same robot on a screen and will follow its gaze,
mimic its behavior, and be more willing to take the physical robot's
advice. We speak more to embodied robots, smile more, and are more
likely to want to interact with them again. People are more willing to
obey orders from a physical robot than a computer. When left alone in
a room and given the opportunity to cheat on a game, people cheat less
when a robot is with them. And children learn more from working with
a robot compared to the same character on a screen. We are better at
recognizing a robot's emotional cues and empathize more with physical
robots. When researchers told children to put a robot in a closet (while
the robot protested and said it was afraid of the dark), many of the kids
were hesitant. Even adults will hesitate to switch off or hit a robot, especially when they perceive it as intelligent. People are polite to robots
and try to help them. People greet robots even if no greeting is required
and are friendlier if a robot greets them first. People reciprocate when
robots help them. And, like the socially inept Clippy, when people don't
like a robot, they will call it names.

What's noteworthy in the context of human comparison is that
the robots don't need to look anything like humans for this to happen.
In fact, even very simple robots, when they move around with "purpose," elicit an inordinate amount of projection from the humans they
encounter. Take robot vacuum cleaners. By 2004, a million of them had

been deployed and were sweeping through people's homes, vacuuming dirt, entertaining cats, and occasionally getting stuck in shag rugs. The first versions of the disc-shaped devices had sensors to detect things like steep drop-offs, but for the most part they just bumbled around randomly, changing direction whenever they hit a wall or a chair.

iRobot, the company that makes the most popular version (the Roomba) soon noticed that their customers would send their vacuum cleaners in for repair with names (Dustin Bieber being one of my favorites). Some Roomba owners would talk about their robot as though it were a pet. People who sent in malfunctioning devices would complain about the company's generous policy to offer them a brand-new replacement, demanding that they instead fix "Meryl Sweep" and send her back. The fact that the Roombas roamed around on their own lent them a social presence that people's traditional, handheld vacuum cleaners lacked. People decorated them, talked to them, and felt bad for them when they got tangled in the curtains.

Tech journalists reported on the Roomba's effect, calling robovacs "the new pet craze." A 2007 study found that many people had a social relationship with their Roombas and would describe them in terms that evoked people or animals. Today, over 80 percent of Roombas have names. I don't have access to naming statistics for the handheld Dyson vacuum cleaner, but I'm pretty sure the number is lower.

Robots are entering our lives in many shapes and forms, and even some of the most simple or mechanical robots can prompt a visceral response. And the design of robots isn't likely to shift away from causing these biological reactions—especially because some robots are designed to mimic lifelike movement on purpose.

BIOLOGICALLY INSPIRED DESIGN

In 2019, I visited Boston Dynamics. The US-based company, currently part of the Japanese conglomerate SoftBank, was founded by roboticist Marc Raibert and spun out of MIT in the 1990s. The company became one of the world leaders in robot locomotion and is best known for developing biologically inspired robots, some of them four-legged and eerily animal-like. The company became famous even outside the robotics world through its popular YouTube videos of robots that could walk around and keep their balance in tricky situations. The company, however, didn't interact much with the press, giving it an air of secrecy and leading to some speculation as to its goals, what parts of their videos were accurate, and what else was going on inside its walls. When I arrived at their unassuming-looking office complex in Waltham, Massachusetts, I was expecting high security and to be asked to sign an NDA. Instead, Raibert greeted me in person and spent over two hours showing me around.

As we were touring the facility, Raibert and I walked past a bipedal robot that was slumped over on the floor of a room. I joked to him that the scene looked just like the robotics labs at MIT, where people envision us working with buzzing, functioning robots all day when really our machines are rarely in operation (and frequently in need of repair). Raibert looked at me and wordlessly opened a door to the next room. My jaw dropped. Behind the door was a gymnasium-sized hall outfitted with an elaborate obstacle course. Dozens of dog-sized robots were roaming the premises, walking up and down stairs, pacing back and forth in pens, or ambling around the area completely by themselves. Out of politeness, I left my phone camera in my purse, but I will never forget the visceral feeling of standing there, watching the zoo-like spectacle of so many completely autonomous, animal-like robots wandering around the room at once.

To many observers, the lifelike robots that Boston Dynamics creates are intriguing, thrilling, scary, or magical. But creating magic and thrills is the goal of places like Walt Disney World. Despite putting out some popular, amateur-produced YouTube videos of their engineering progress, entertainment is not the objective of companies like Boston Dynamics. So why do they create robots that move like animals?

Around 350 BCE, Greek philosopher and mathematician Archytas of Tarentum created a flying wooden pigeon. Powered by steam or compressed air, the dove could allegedly fly for hundreds of feet before plummeting to the ground. Archytas, whom many have deemed the founder of mathematical mechanics, created with his pigeon both one of the first "robots" and one of the first studies of flight.

Over two millennia after Archytas's pigeon, we're still building mechanical birds to understand how things fly. For example, in 2020, researchers at Stanford University made a robot called PigeonBot. Unlike Archytas and many others who attempted to create a physical simulation of bird flight, they outfitted their robot with real feathers. By observing their PigeonBot, they discovered a property in feathers called directional Velcro. The "Velcro" lets the feathers slide around to change the shape of the wing but lock together whenever the movement creates holes or spaces in the wing surface, making it a key mechanism for controlling flight.

Biomimetics (also called biomimicry), is a broad academic trend of looking to solutions in nature in order to problem-solve. A large subfield of biomimetics focuses on biologically inspired robotic design, in particular animal-inspired sensing, body design, and locomotion. Animals crawl, run, fly, swim, jump, swarm, and climb in a myriad of different ways, all of them well suited to the environments they thrive in. Basically, if we're trying to make robots that need to propel themselves forward or navigate natural terrain, it makes sense to explore what biological movements have evolved in nature and try to apply those to robotic systems.

Research labs around the world are creating robots that move like animals, from robotic cheetahs to turtle-bots to autonomous metal cockroaches. Swarm roboticists study the collective behavior of animals like insects or fish in order to create robots that communicate, move, and work together as a group. Some robots are built to be soft and malleable, making them able to propel themselves forward like worms or snakes.

As we design robots, we sometimes end up going in different directions than biology has taken us. After all, despite Leonardo da Vinci's bat-inspired flying machine, we didn't wind up with the airplane until we explored an aerodynamic form that deviated from nature. But as in the case of feathered PigeonBot, the diversity of solutions in nature is a great starting point for us to learn some of what works and what doesn't. Because it's so downright practical to borrow movement from nature, we now frequently encounter biologically inspired robot design in research and development, and some of it will probably become common in the machines around us.

One of Boston Dynamics' first famous creations, BigDog, was a mechanical pack mule. It could carry up to 340 pounds and was designed to go where vehicles couldn't, lumbering over all sorts of terrain without

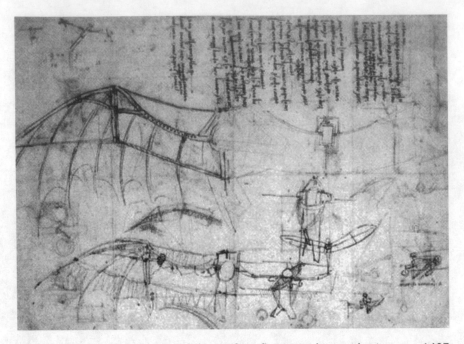

Leonardo da Vinci's bat-inspired design for a flying machine with wings, ca. 1487

losing its footing. It was ultimately too loud to be practical in most settings, but the technology has improved. Nearly all of the Boston Dynamics robots are biologically inspired in much more than just their names—it's actively difficult to watch LittleDog, Cheetah, or their commercially available Spot stumble around and not feel, on a deep level, that they are more animal than object.

Robots that feel alive aren't going away. In fact, this situation is only going to get more intense. While evolution created self-propelling animals, technology lets us advance robotic design more quickly than nature. It also lets us be more intentional about what we're doing.

Beyond the practicality aspect of robot mules, there's another reason to design robots that move in a lifelike way—it lets them socialize with us. When my son was two and a half, we were using a variety of home assistants in our household: Apple's Siri on our phones, an Amazon Echo Show in our kitchen, a Google Home in the bedroom, and a robot named Jibo in our living room. Jibo was the brainchild of MIT professor Cynthia Breazeal, whose goal was to create a social home

BigDog (2012)

robot. Unlike our more static speaker assistants, Jibo had a body that could move. It swiveled its head in the direction of people and conversations, made "eye contact," and could perform a variety of dances from slow to twerking. The tech press gushed: "Jibo isn't an appliance, it's a companion, one that can interact and react with its human owners in ways that delight." When the robot was first announced, the small startup bringing it to market received a slew of positive attention and $25 million in series A investments.

Of all the interactions we had with these personified assistants on a daily basis, my child only cared about Jibo. He would say hi to Jibo and ask us (because Jibo didn't understand his toddler pronunciation) to make the robot do things, like recite a poem, do a yoga sequence, or scan the room for monsters (for which Jibo would spin in a circle and then reassuringly announce that the room was monster-free). My son understood and responded to Jibo's movement and social cues. When my son was learning the word "robot," he identified Jibo and everything from robot-shaped figurines to our Roomba, but it never occurred to him to include the other virtual assistants.

Like Jibo, some robots are intentionally designed to tap directly into our biological social and emotional responses. These robots, which communicate life and agency in order to socially interact with people, are called social robots. And they work on a deep level. According to psychologist Sherry Turkle, "When robots make eye contact, recognize faces, mirror human gestures, they push our Darwinian buttons, exhibiting the kind of behavior people associate with sentience, intentions, and emotions." As robots move into shared spaces, we're going to see more social robots that are specifically designed to interact this way. But there's a common misconception that these robots will be designed to be as much like people as possible, when, contrary to popular belief, these social robots don't need to look anything like humans.

SOCIAL ROBOT DESIGN: LESS IS MORE

Entertainment automata are much older than the word "robot." The ancient Greeks used automated mannequins in ceremonies and theater. In 1495, Leonardo da Vinci built a mechanical knight that could move its arms and raise its visor. Sixteenth-century clockwork technology inspired a slew of mechanical creations, like a fifteen-inch-tall monk made of wood and metal that could walk around, nod its head, and raise the cross in its left hand to its lips. It also kicked off Japan's big trend of Karakuri puppets (からくり人形), traditional Japanese mechanized dolls, which began sometime during the early seventeenth century.

More advanced than my childhood memories of clunky metal automata playing "Happy Birthday" at Chuck E. Cheese (sadly never for my own birthday), the impressive moving animatronics that fill Walt Disney theme parks are so compelling that they wow roboticists, even when they don't fit the definition of a smart machine. At a recent conference, most of the talks were about impressive breakthroughs in artificial intelligence and robotics, but the final presenter got more reactions from the audience of hard-core scientists than every other demo combined. He was from Walt Disney Imagineering, the entity that leads the research and development of Disney's theme park attractions. The videos he showed were of automata that had the entire room breathing "whoa" in unison.

A lot of the robots we've created for art and entertainment purposes throughout history are designed to look like humans. But it's not necessary for robots to look human to engage us. Our fascination for robots is paralleled by our fascination for animals, which we've also had perform for us throughout history. We've put them on display and trained them to do entertaining tricks. Lions were kept in cages thousands of years BCE. In ancient Roman arenas, animals fought each other (and people) for show. Traveling collections of animals delighted small towns and made money for their owners. Today, despite increasing opposition from animal welfare groups, people still gather in droves to enjoy circus performances, dolphins and seals doing tricks in aquarium shows, and piglet races at county fairs. Wildlife theme parks are still immensely popular, as are zoos.

Even though some roboticists claim that the ideal social robot looks and behaves just like a human, successful social robot design often goes counter to this claim. The trick to getting our attention isn't necessarily to look like us—it's simply to mimic cues that we recognize and respond to.

Cynthia Breazeal's love for the characters from *Star Wars* led her to pioneer the field of modern social robotics. As a doctoral student at MIT, Breazeal did research in the 1990s on social expression between humans and robots, developing one of the first social robots: a large, mechanical face with lips and furry eyebrows named Kismet that was designed to recognize and express emotion.

Only a small percentage of our communication happens through our words. Most of it is nonverbal—transmitted through body language, tone, gesture, and facial expression. The characters we're drawn to don't need to have human language. For example, it's possible to express emotion so that people can recognize it solely through our movements, like dancers. Heather Knight, a former Breazeal student, directs the CHARISMA Robotics Lab at Oregon State University, a lab that creates and studies minimal and "non-anthropomorphic" social robots. For example, her group has created a robotic chair that approaches people

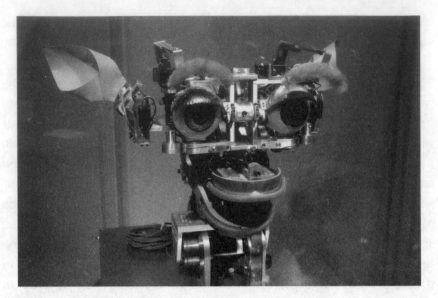

Kismet at the MIT Museum in 2013

and tries to persuade them solely through its movements to stop by an outdoor café for refreshments. Another former Breazeal student, Guy Hoffman, has used his background in animation and acting to design non-humanoid robots, for example, a desk lamp that expresses emotion through motion.

These robots engage us because we're drawn to the recognizable human cues in their behavior. For the past century, animators have honed the art of taking human cues and putting them in animals or objects to create something even *better* than people. Both Disney and Pixar have a history of crafting adorable creatures and objects with a range of emotional expressions, from dinosaurs to teapots. Today, the social roboticists I work with often hire professional animators to help design their machines. Even though the current technology is in its infancy, it works. It turns out that robots don't actually need to look as realistic as Data from the show *Star Trek: The Next Generation* for us to anthropomorphize them. In fact, they're often more compelling to us when they are different.

Aside from tapping into our inherent preference for something cuter than real life, for example, exaggerated, baby-like faces, there's another

reason not to aim for human replicas: the "uncanny valley" (bukimi* no
tani genshō (不気味の谷現象). In the 1970s, Masahiro Mori, a professor of engineering at the Tokyo Institute of Technology, came up with
the idea that we increasingly enjoy robots the more humanlike they
become, but as the design gets too close to humans without replicating
them exactly, our affinity for them changes to revulsion. Mori's theory
also explains, for example, why we think zombies and prosthetic hands
are creepy, but not (for the most part) other people or stuffed teddy
bears.

Not everyone agrees with the hypothesis. The valley itself has been
tested empirically, and the results have been mixed. But Mori's concept
persists and continues to resonate intuitively. One of the areas where it

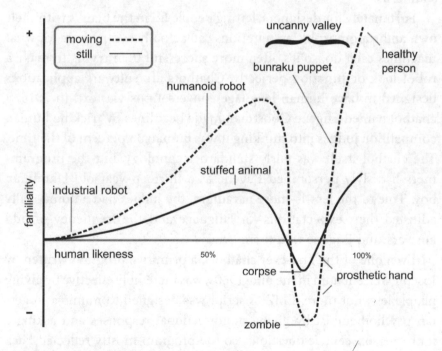

The uncanny valley hypothesis, according to Mori

---

* Bunraku is a form of traditional Japanese puppet theater from the seventeenth century.

bears out in some form is in the social robotics design principle of expectation management. When we see something that closely mimics a person or an animal that we're intimately familiar with, our brains expect it to behave exactly like its counterpart. When it inevitably makes a strange movement or otherwise doesn't quite live up to that expectation, we get thrown off. This experience is disorienting, making us feel insecure in our predictions, and could explain the "creepy" factor that some people experience with very humanlike robots.

In robotics, expectation management is important. Would people like their Roomba better if it was a humanoid that walked around vacuuming? Probably not. Anything that looks human raises people's expectations for how it performs and what it should be able to do, leading to quick disappointment when the robot doesn't live up to our standards.

Fortunately for designers, letting people fill in the blanks with their own anthropomorphic assumptions can actually compensate for what machines can't do, so it's often more successful than trying to make a robot look or function perfectly. Chatbots are software applications designed to have human language conversations via text. In 2014, a chatbot named Eugene Goostman made headlines by tricking human competition judges into thinking it was human 33 percent of the time. The chatbot itself was fairly standard technology, but the programmers had slyly personified Eugene as a thirteen-year-old Ukrainian boy. True to the less-is-more paradigm, the judges had automatically adjusted their expectations for language and the frequency of odd answers, and Eugene won them over.

Even one of the first-ever chatbots, a primitive program written by Joseph Weizenbaum in the mid-1960s, was incredibly effective by giving people less, not more. ELIZA's script was designed to mimic a Rogerian psychotherapist by giving nondirectional responses and asking a lot of person-centric questions, so the program mostly reflected back any statements that people gave it. For example, if someone typed *"I feel depressed"* into the input field, ELIZA might ask them why they felt depressed or respond with something open-ended like *"Tell me more about your feelings."* Because ELIZA would prompt instead of responding in kind, the computer program didn't have to say anything smart.

People love to talk about themselves, and the people interacting with ELIZA would suspend their disbelief and chat away.

The idea of letting people fill in the blanks themselves applies to much more than just human language. A lot of today's social robots use a screen to display a face. But even though this would technically allow for very detailed and realistic animations, the designers usually opt only for a few simple elements. In a 2018 survey of screen-rendered robot faces, less than half had eyebrows and less than a quarter of them had a nose. Doing more would be counterproductive because our anthropomorphism, when faced with a lack of information, happily fills in the rest. The same is true of behavior. Communication through simple cues that don't require navigating the complex world of human speech, like movements, blinking lights, or beeping, is often more effective, and less disappointing, than words. And even introducing some intentional "mistakes" like randomness and unpredictable behavior can increase people's anthropomorphism toward a robot.

When I unboxed the newest, modernized version of Sony's AIBO in 2020, the instructions informed me that the robotic puppy wouldn't always follow commands because it was "moody" and had "a mind of its own." I don't know whether the robot is intentionally designed to behave randomly or whether Sony's description is mostly there to paper over the occasional malfunction, like when the robot doesn't hear a command. But it doesn't matter. We named it Analog, and my husband talks to it as though it were a real puppy.

Research in the rapidly growing field of human-robot interaction confirms that we respond strongly to social robots. And we enjoy doing it. We know intellectually that these robots are just machines, but we're both willing and eager to give them a relational role.

As social robots increasingly enter our world and get more attention in the media, the most frequent concern that I encounter about them is the worry that they will replace human relationships. It's clear from the research that social robots don't need to look or function like humans to be liked (and are in fact often more likable when they differ). But the other thing to understand is that this isn't the first time we've navigated social relationships with nonhumans.

We've long used animals for work and weaponry, but they have also become our friends. And we're going to start sorting robots into different buckets like tools, products, and companions, just like we've done with animals. This is because our relationships with animals are driven by the same impulse that forges human-robot relationships—our tendency to anthropomorphize.

5

(HU)MAN'S BEST FRIEND

THE HISTORY OF COMPANION ANIMALS

[Content warning for this chapter: animal suffering]

"I sometimes think that, in the desperate straits of humanity today, we would be grateful to have nonhuman friends, even if they are only the friends we build ourselves."

—Isaac Asimov, *Robot Visions*

In World War II, some countries asked their citizens for more than just human recruits—they also asked for their dogs. A 1930s rally in Germany urged German families to donate over 16,000 personal pets to the military's canine corps. America followed suit in the 1940s, when an organization called Dogs for Defense ran radio advertisements and posters to convince tens of thousands of Americans to send their puppies to war. So many pets were donated in response to the "dog soldier" recruitments that the number of dog shows in America plummeted. Dogs went from chasing neighborhood squirrels to being spies, guards, and messengers in the war. But the armed forces didn't fully anticipate what this would mean.

The US military suddenly found itself inundated with letters from the owners asking about the pets that they had sent in. According to Susan Orlean, author of *Rin Tin Tin: The Life and the Legend*, "Just as people wanted to know how their friends and family were getting along in the army, they wanted to know how their dogs were doing. They sent them Christmas cards and birthday cards and wrote to the army, asking after Butch and Chips and Peppy and Smokey." The military wasn't equipped to deal with people's demand for information about their beloved pups. They sent out basic form letters to try to appease the dog owners and created propaganda stories about happy dogs reuniting with their families after the war. Still, people complained.

The effort to create a volunteer American dog corps in World War II, while otherwise successful, was completely unsustainable from an emotional labor perspective. After this experience, US military efforts continued to involve canines, but they were no longer solicited from people's homes, and they were treated strictly as property. What made the American war dogs of the 1940s so different was their status as pets. The role of dogs had recently started shifting from guarding people's yards to living in their homes, and they were becoming part of the American family.

While a lot of our animal partnerships involve using them as tools, we've also developed social relationships with the autonomous beasts in our lives. Especially in recent decades, we've treated certain animals very differently than others and given them a place in society as our companions, family members, and loved ones. But our tendency to bond with animals goes back much further in our history, and illustrates the depth and diversity of our social relationships, as well as why we create them.

THE HISTORY OF PETS

In 1926, some generous admirers sent US president Calvin Coolidge a meaty dish for his family's Thanksgiving dinner: a live raccoon. The animal-loving president and his wife refused to eat the furry, beady-eyed creature and kept it as a pet, instead, naming it Rebecca. Soon, Rebecca was wearing a beautifully embroidered collar with the label "White House Raccoon" and joining the president for daily walks, quickly becoming a darling of the public. Raccoons, while cute, are not commonly kept as pets, and for good reason—they're mostly untamable, not very social, and can be aggressive. To the annoyance of the White House personnel, Rebecca was no exception, which meant she would run away, climb trees, and probably bite the exhausted staffers who spent their days chasing after her. But the family loved her. They built her a house and took her with them on vacation. First Lady Coolidge later reminisced that Rebecca loved "playing in a partly filled bathtub with a cake of soap."

What is a pet? *Merriam-Webster* defines "pet" as "a domesticated animal kept for pleasure rather than utility." Whether or not Rebecca the raccoon was truly domesticated, she certainly wasn't kept for her usefulness. Beyond keeping animals in zoos and circuses for our amusement, they've also tapped into our social nature.

The oldest depiction of a guinea pig in English art is a painting from around 1580, during the reign of Queen Elizabeth I. A seven-year-old girl holds the guinea pig and a five-year-old boy holds a pet goldfinch. Guinea pigs were just beginning to be brought to Europe from South America by the Spanish and were popular as royal pets. But even in

First Lady Grace Coolidge holds Rebecca, the family's pet raccoon, at the White House (1927)

1580, keeping animals as companions wasn't new. The ancient Greek word *athurma*, meaning "plaything," "toy" causing "joy" and "delight," was used to refer to pets.

The roots of our animal companionship likely go back tens of thousands of years to the domestication of dogs in Central Asia. There's a lot of speculation as to why or when dogs appeared. Modern dogs descended from wolves, possibly beginning a transition to "dog" as early as 40,000 years ago, and while the exact details are fuzzy, researchers agree that dogs are the first domesticated animal in history. We probably began a relationship with dogs before we could write or read, before we had agriculture, and before we domesticated other animals like cows or pigs for food.

Experts have different theories as to how they became our first (and "best") friend. Some think that we started to use dogs to help with hunting because they were able to track down prey and could assist with capture. Others argue that we entered into more of a quid pro quo partnership with the canines, joining forces to find food and shelter as we roamed Eurasia together, and coevolving as cooperative species. While we certainly partnered with dogs because their hunting skills complemented our own, over time, they took on a more complex job.

The dog is a good example of an animal species that holds many different tool- or companion-like roles across times and cultures. The persistence of a long-standing dog meat trade in Southeast Asia illustrates that dogs still have a variety of functions today, including being a food source. Long before European settlers arrived in America, bringing with them the precursor to our Western idea of "pets," Native Americans had developed their own human-canine relationships. Dogs took on different roles across Indigenous cultures: canines were herders, hunters, spiritual symbols and sacrifices, food, and also beloved companions.

For millennia, many human societies and cultures that have harnessed dogs as workers have also made friends with them. We know this from archaeological evidence that demonstrates that, even thousands of years before collies and retrievers became part of our families, people mourned their dogs when they died.

The first domestic animal that we know of with a name was

Abuwtiyuw. The royal guard dog in the Giza Necropolis received a lavish burial, the likes of which were normally reserved for upper-class humans. According to the stone inscription, dated around 2280 BCE, Abuwtiyuw was *"the guard of His Majesty . . . His Majesty ordered that he be buried (ceremonially), that he be given a coffin from the royal treasury, fine linen in great quantity, (and) incense. His Majesty (also) gave perfumed ointment, and (ordered) that a tomb be built for him by the gangs of masons. His Majesty did this for him in order that he (the dog) might be honored (before the great god, Anubis)."* In Egypt, killing someone's dog was a capital crime. When a family dog passed away, the family would grieve the same way they would for a person: by shaving all the hair off their own bodies (including their eyebrows). They also believed they would meet their canine companions again in the afterlife.

Cats, which originally came out of the Far East, appear to have a similar relationship trajectory, starting out as useful mouse-catching tools in societies that relied on agriculture. But they also began receiving proper burials as early as 9,500 years ago. Cats enjoyed a particularly revered status in ancient Egypt, where they were admired as intelligent and mysterious. The ancient Greek historian Herodotus claimed that the Egyptian people would have saved a cat from a fire before anything else, even themselves. Killing them was punishable by death, and an entire branch of government was devoted to preventing the export of cats.

The types of animals we relate to aren't just our original hunting partners. Similar to the diverse range of animals we've used for work, our companion animals have also spanned a variety of different species. Alongside their beloved cats and dogs, the Egyptians had gazelles, monkeys, and all sorts of other animals as pets. They kept birds as early as 4000 BCE. Central Asian societies housed horses and prized them as far more than just work animals. Some hunter-gatherer societies kept pets, such as monkeys.

By 1870, we pretty much had the same range of animals as pets that we do today, with a few cultural differences. For example, in modern Japan, fish, in particular koi, are more popular than in other countries. The Niigata area of the west coast of Japan started breeding carp with beautiful coloring around 1800, but they didn't explode in popularity until after the 1914 Great Tokyo Exhibition. By 1996, half of all Japanese pet owners had goldfish or koi (compared to 12 percent of American households having fish of any sort at all).

Why do we keep pets? During much of history, pets were status symbols of the rich. The East Asian, Aztec, and Western European elites loved small dogs. Around the 700s, the Pekingese became a palace dog, and Chinese emperors would have anyone who stole one executed. The Spanish and Portuguese started capturing canaries in the Canary Islands and selling them to wealthy Europeans in the 1500s.

Henry III of France received ambassadors while carrying his favorite dogs in a basket hanging around his neck. In the 1700s, Frederick the Great of Prussia became obsessed with miniature Italian greyhounds. He owned thirty-five of them over the course of his life and gave them their own chairs covered in silk. But while pets were a display of privilege, the elites also showed considerable emotional attachment toward their status symbols. When one of Frederick's dogs was captured during the Seven Years' War, he negotiated its release through a proper prisoner of war exchange. European royalty had their portraits painted with their pups and even started having clothes made with special pockets into which they could stuff their tiny dogs, so that they could keep them close. During the Victorian era, Queen Victoria owned eighty-eight pets. Cat and dog shows started to become popular and, later in the century, animal lover clubs. Owners so valued their dogs that they were commonly kidnapped for ransom or reward.

Pets haven't always been welcomed everywhere. During the Chinese

Catherine of Aragon with her pet monkey
(Horenbout, ca. 1525–26)

Communist era, dogs were utilized as workers or food, and pets were largely considered a bourgeois luxury. In 1983, the city of Beijing flat-out banned dog ownership. But some people wanted to have dogs so badly that they pleaded for a change in rules (or simply didn't follow them), so in 2003, Beijing finally relaxed the ban and began to allow small dogs that were no taller than 35 centimeters (~14 inches). Today, pampered little dogs abound on Beijing's streets, often seen taking walks with their owners in the early morning.

The dog trend in China follows those of other countries with a rising middle class like Japan and those in Western Europe. After the Second World War in the United States, the middle class experienced an economic boom. People started gravitating toward a new, culturally established middle-class norm of a suburban house, two children, and a dog. Pets became an investment in entertainment and companionship for the kids.

Unlike our wolf-descended partners of the past, there are now many animals in our lives that we don't even pretend to keep as guards, hunters, or helpers. They are in our homes for our enjoyment and companionship. And that relationship is evolving, too: we give our pets more autonomy and agency than we used to (rather than strictly expecting them to obey), and we spend an increasing amount of time and money on their well-being.

According to the American Pet Products Association, which conducts a national pet owners survey every year, 56 percent of US households owned a pet in 1988. Today, in 2020, pet ownership exists in 67 percent of households (totaling a whopping 84.9 million homes). But while pet ownership has grown, the bigger change has been in how much time, money, and emotional energy people are spending on their animal companions. In the United States, the pet industry was valued at $28.5 billion in 2001. It more than doubled by 2016, growing to $66.75 billion. Now, in 2020, the industry is approaching a worth of $100 billion.

It's easy to explain why people might domesticate cows, pigs, and chickens—those animals are useful to us. But what drives us to want to spend so much money on a Chihuahua? Some of it is love. And yet, just like with robots, that love may be more self-centered than we like to admit.

ANTHROPOMORPHISM (PROJECTING OURSELVES)

Dogs used to be a traditional menu item in Beijing, but now that the pet dog ban has been lifted, there are more dog-treats than dog meat available for consumption. Beijing also has dog social networks, dog swimming pools, pet spas, and grooming establishments that will cut and dye your fluffy friend to look like a panda or a Ninja Turtle. The $100 billion that Americans spend on food, supplies, and services for their animal friends doesn't even include the sale of dogs, cats, or horses, which are also only a small part of what the pet industry peddles.

The Barkley Pet Hotel & Day Spa, with locations in the US and India, is one of many luxury resort-style vacation facilities for pets. Dogs can be booked in "executive poolside suites" complete with artwork, designer bedding, in-suite flat-screen TVs with animal programming, and webcams so that their owners can call them 24/7. Relaxation packages for felines feature organic catnip and sushi. The room service menu includes tender cut, chargrilled filet mignon served with a fine Irish linen napkin. During the day, the animals are treated to arts and crafts, pool parties in heated saltwater pools, ice cream socials, aroma-scented relaxation massages, and even an excursion by car to the McDonald's drive-through to get a hamburger, cheeseburger, or Chicken McNugget snack. And, according to the Barkley's website, "A nighttime tuck-in tummy rub and a refreshing spritz of bottled water provide the perfect ending to each day."

While the pets no doubt enjoy at least some parts of this lavish treatment, a lot of these services cater to our tendency to project our own desires onto our animals. Data from the American Pet Products Association shows a consistent increase in companion animal services that cater directly to people's anthropomorphism: luxury pet products now include gourmet foods, designer clothing, nail polish, edible birthday cards, breath fresheners, and designer label shampoos and perfumes. An airline named Pet Airways, launched in 2009, promised "World Class Air Travel for Pets," placing dogs and cats (which they called pawsengers) in the main cabin instead of the cargo hold for their journey. People get their pets holiday gifts, send them to yoga and acupuncture sessions, and pay for their own pregnancy photo shoots and weddings.

The pet product and service industry is well aware of our tendency to treat our pets like mini-humans, and as companies capitalize on it, they may also be perpetuating it. But anthropomorphism toward our pets isn't a twenty-first-century invention. In 1890s Paris, people would dress up their pet dogs in coats or bathing suits, and even buy them fancy underwear. Zooarchaeological evidence tells us that the ancient Romans spoiled some of their dogs royally. Unearthed remains indicate that there were multiple smaller "toy" breeds of dogs in Rome, many of which show signs of excessive human pampering and feeding.

We've also bombarded children with anthropomorphic portrayals of animals for quite some time now. In 1860s America, nearly one hundred years before Walt Disney made the popular films *Lady and the Tramp* and *One Hundred and One Dalmatians*, children's literature started depicting animals as versions of ourselves. Beatrix Potter published the trade version of her iconic book *The Tale of Peter Rabbit* in 1902. Kids' books increasingly placed animals at the center of the story, creating furry protagonists that had names, wore hats and pants, danced, sipped tea, and learned moral lessons.

This enthusiasm for viewing animals as creatures with human desires and behaviors wasn't just for children. Around the same time, the printed trade cards that people loved to collect included lots of pet animals posed as people: eating at tables, wearing dresses, and pushing baby carriages. Popular writer Reverend William J. Long published multiple books about the animals that lived around his home, giving them exaggerated anthropomorphic characteristics. Readers gobbled up his claims that a raccoon buried stolen chicken because it felt guilty or that animals would heal their own broken bones by making little casts out of clay. People's human-inspired view of animal behavior also led to scams like Clever Hans, a horse that people were eager to believe could do math calculations and tap out the answers with his hoof.

Already popular in literature, pets as anthropomorphic characters experienced another surge in popularity with the arrival of the silver screen. The dogs that were made the stars of silent movies were often portrayed as having humanlike intelligence, bravery, and emotion. In the 1914 film *A Dog's Love,* a mourning pup purchases flowers, brings them to a grave at a cemetery, and even convinces a passerby to water

One of the "smart set" (ca. 1906)

them. After World War I, German shepherd movie stars Strongheart and Rin Tin Tin became famous, appearing on cereal boxes, dog products, and in magazines. Their film characters were sentimental and heroic: the dogs showed sorrow, bravely rescued people from danger, and performed feats of humanlike intelligence like lighting a lighthouse lamp to save a ship from the rocks.

After the Second World War, as dogs moved into people's homes and families, the representation of dogs in media shifted to a more intimate role. The next big canine movie legend was a collie named Lassie. In the 1943 feature film *Lassie Come Home*, the character won the hearts of mainstream moviegoers and began her long history as an iconic American dog. Lassie helped create a new image of dogs as lovable. Her portrayal in movies, radio, and television as faithful, loyal, and compassionate likely influenced many people's relationship with dogs in modern Western society. But the anthropomorphic depiction of "perfect" dog Lassie also set a high bar. According to behaviorist and zoologist Patricia McConnell, author of *For the Love of a Dog*, Lassie the character led people to believe that "good" dogs didn't need to be trained to do things, could understand all human language, and had an inherent desire to obey and please their owners.

Our penchant for projecting humanlike qualities onto our companion animals becomes especially apparent when we get it wrong. Just like with robots, when we lack knowledge of how or why an animal behaves in certain ways, our anthropomorphism tends to fill in the blanks. We are prone to make a lot of untrue assumptions about our furry companions, and not just about whether they will enjoy their room service on a fine Irish linen napkin.

Studies have shown that people, even though they view dogs as different from children, will falsely attribute similar types of mentalities to both dogs and children. In a study that looked at whether dogs would go get help if their owners were in trouble, a skill they're commonly believed to have, not a single one of the pets lifted a paw to save their human significant others. Another common assumption is that dogs will feel guilty if they have "been bad." When you walk into your living room to find your favorite slippers shredded to pieces, your dog may have what you believe to be a distinct look of guilt on her face. But a study showed that dog owners get it wrong: the look on their pup's face may simply be responding to an owner's upset facial expression rather than an admission of bad behavior.

Despite plenty of research and knowledge that contradicts our storytelling and intuition, we humanize our pets: we name them, talk to them, buy them rhinestone collars and fancy food, and we eulogize them like people when they pass away. In fact, it's nearly impossible for us to stop ourselves from anthropomorphizing animals, pets or otherwise. We enjoy looking for emotions and behaviors we think we recognize, even if they're not there, which also explains people's love for their robot dogs. According to some experts, our desire to project our own feelings and see them in others is the main reason we have pets at all.

Of course, when it comes to animals, our assumptions aren't always wrong: animals have experiences of the world that we can relate to. For example, dogs *do* have feelings and they *are* smart. They enjoy and dislike things and can disobey commands. But it becomes clear that our pets' inner worlds matter less to us than we think when we see how people treat their robots, and even when we look at how we treat some of our living pets.

While it may seem mostly harmless to pretend our companion animals are little people and surprise them with Frosty Paws ice cream, there are cases where our anthropomorphism leads us further astray. In late 2017, veterinary cardiologist Dr. Anna Gelzer and other vets noticed a sharp increase in dog heart disease. When they investigated, they discovered a link between the severe heart conditions and grain-free dog foods. Grain-free dog food uses alternatives to grains, like lentils or chickpeas. It used to be a rarity, only available for pets with allergies. But by 2017, it had become the hottest new trend in pet foods. Even though there was no scientific evidence that it was good for the animals, a lot of people were putting their pets on grain-free diets. Why?

The trend in grain-free dog food happened in parallel to the same trend in human diets. A decade ago, Americans started demanding that their food be low carb and gluten free. Lisa Lippman, the lead New York veterinarian for the in-home veterinary service Fuzzy Pet Health, was not surprised: "People will anthropomorphize or project onto their pets whatever they think they need to eat themselves." Even though pet owners want what's best for their furry friends (or perhaps because of it) veterinarians have struggled to persuade dog owners to switch back to grain and stave off the heart problems in the dogs. According to Christopher Lea, director of the Auburn University Veterinary Clinic, people remain compelled to foist their own food preferences and anxieties onto their pets by controlling their dogs' diets.

The only reason the vets figured out what was happening is that they started seeing an increase in smaller dogs with heart problems. Large dogs wouldn't have raised a red flag because so many of them are already prone to heart conditions—due to breeding. In the context of comparing animals and robots, it's interesting to note that we've also "designed" many of our pets.

BREEDING (DESIGN)

I remember distinctly the day my pet hamster died. We buried her in a box in the yard in a formal-ish ceremony while I, about eight years old, cried a few tears. The hamster is one of the most common small animal pets in family households today. The furry little rodent hit pet stores around 1945, after being captured near Aleppo in Syria and bred in a university lab in Jerusalem. Like many other children of my generation, my brother and I each received one of these short-lived, docile, and adorable companions.

I loved my hamster. She was plump and fluffy, with a little nose and big black eyes, all attractive features that humans are drawn to. But one of the reasons hamsters have a short life span (making them an easy time investment for parents) is that they are prone to terrible health conditions. Their overbreeding causes them to perish from congestive heart failure, kidney disease, and other incurable ailments.

While roboticists are working with animators to design social robots that appeal to us, we actually have a longer history of designing our animals. But, unlike animating machines, shaping animals is a long road full of hardships—for the animals. Breeding horrifyingly illustrates some of the self-serving nature of our animal relationships. Despite the fact that controlled reproduction causes a myriad of inherited problems in our pet animals, we have a history of breeding them to encourage certain traits, not the least of which is their appearance.

In the 1950s, one of the biggest US fashion trends was a dog. People started going wild for poodles, and they were appearing everywhere: poodle skirts stitched with felt poodles, poodle purses, rhinestone poodle barrettes, poodle wallpaper, and, of course, real poodles. The curly haired breed was regarded as the embodiment of sophistication and became the hottest new status symbol. People bought poodles, flaunted them, took them to expensive groomers, and even dyed their fur with vegetable dye to match their clothes. By the 1960s, the poodle had become the most popular breed of dog in America.

Today, the poodle is less popular than breeds like retrievers and bulldogs. But while their heyday of celebrity photo shoots is over, there is still a subculture that practices turning poodles into works of art. Creative dog groomers specialize in sculpting and coloring poodle fur to shape the dogs into bright canvases of flowers, dinosaurs, mermaids, and more. It renders the dogs so unrecognizable that my friend's dog-loving toddler didn't even realize that the subjects of the poodle-grooming documentary *Well Groomed* were her favorite animal.

Breeding is when we reproduce an animal species within a smaller population to encourage specific, desirable traits. It's something that can happen naturally in the wild, and we've also done it very intentionally, for example, to make our domesticated animals stronger workers or better steaks. The modern, broad-breasted turkeys we use for meat are bred for flavor, magnitude, and developmental speed, growing to full size in four months and ballooning up three times larger than wild turkeys. But we've also gone from domesticating wolves to be fierce hunters and guard dogs to breeding curly haired fur balls that we name Princess and sculpt to look like Ronald McDonald.

It's not quite clear when we first started to experiment with pet breeding. There's evidence that humans intentionally bred sled dogs in remote Siberia as early as 9,000 years ago. Japan got into breeding albino and other unusually colored mice in the 1700s, starting a trend that caught on in Europe in the 1800s. But it was during the nineteenth century that people began to keep close track of dog breeds in order to refine them, resulting in what we know as modern dog breeding for pedigree. Pedigree cat breeding wasn't far behind, with cat shows exhibiting a wide variety of different felines starting as early as 1871 in London. By the time Walt Disney released the 1955 animated film *Lady and the Tramp*, cultivating more exotic-seeming breeds like Siamese cats was all the rage in Western society. But cats have always been less variable than dogs in their genetics. In fact, most animal species pale in comparison to the wolf genes that we've been able to shape into so many different canine skills and appearances.

"Designing" our dogs has enabled us to cultivate a variety of traits in our animal partners, but when we select for our pet breeds, it's not always about making them stronger, healthier, or better tempered. Sometimes it's to accommodate our needs, like the Labradoodle. First bred by Australian Wally Conron in 1989, the cross between poodle and

Egyptian dog types

Labrador retriever alleviated some people's dog allergies by selecting for poodle hair instead of fur. And sometimes, we select for fashion at the expense of temper. Poodles, America's darling dog of the 1950s and '60s, are about as aggressive as pit bulls and rottweilers, and just as likely to attack a child. But when it came to their popularity, their personalities mattered less to Americans than their appearance. Oftentimes, instead of selecting for intrinsic traits, we've turned dogs into fads, like our outfits or cars, and bred them to be whatever is fashionable. We've also selected them based on what looks most attractive to us.

Breeding pets for looks can directly follow our anthropomorphic tendencies. In what's called the "cute response," we gravitate toward animals that look most like human babies, with big eyes, small noses, and plump, soft bodies. Animals that look like our own young make us want to take care of them, and we prefer animals that most clearly display emotional cues we recognize to be like ours. "Puppy dog eyes" are more compelling to us than their ancestor wolves' features, so, over the course of fairly recent history, we've intentionally created dogs that better mirror our desirable emotions. Similar to the way we design social robots with big eyes and round shapes, we've tried to "design" the same in our dogs.

But breeding animals for attraction instead of function has been largely detrimental to the health of our pets. By the 1960s, hip dysplasia had become a huge problem and a widespread condition that dogs inherited through intentional breeding, raising concerns among animal medical professionals. It wasn't the only health issue: heavily bred dogs started developing thyroid disorders, heart problems, and other severe diseases. Because we've bred English bulldogs to have giant, babylike heads and even shorter snouts, the features that people are drawn to, it means that most of them have to be born by caesarian and are crippled by breathing issues, many of them dying of chronic oxygen deprivation. And we've done the same with pugs, boxers, and other popular snubnosed breeds.

Thanks to public awareness campaigns on some of the health problems, breeding is becoming increasingly controversial, and the popularity of keeping purebred dogs as pets has decreased. But the dog and cat shows that revolve around purebreds persist worldwide. And the effects of having engaged in this genetic experimentation are massive.

According to anthrozoologist Hal Herzog: "We have jiggered the dog genome into a bewildering array of animals that look magnificent but that, in the end, are like modern tomatoes, a triumph of style over substance."

And yet, we love our pet tomatoes. We may love them most when they look like us and when we can see our own emotions in them, but there's no doubt that we care deeply for them. People will devote massive amounts of their time, money, and energy to the well-being of their animal companions, even going so far as to structure their whole lives around them. Some of my friends have parrots that dictate the time they need to return home every night to "put the birds to bed." Another friend dedicated over a decade of his life to trying to alleviate his rescue dog's anxiety, buying her every type of weighted dog blanket available and coaxing her to eat Prozac before every thunderstorm. And Ira Glass, host of the popular NPR radio show *This American Life*, is known to have spent an extensive amount of time caring for a dog with a myriad of mental and physical health problems, including sacrificing his social life and sourcing different types of kangaroo meat to deal with his pup's food allergies.

We've clearly come a long way from domesticating animals to use them solely for work, weaponry, and other practical purposes. We've developed such a strong emotional connection to certain animals that nobody would deny that we love them. And this love has actually been useful to us: our animal friends are able to benefit and help us specifically *because* we have a socio-emotional connection to them. This is important, because it's a potential benefit of robots, as well, and it's one that's often overshadowed by what I believe are the wrong concerns. But before I get back to robots, let's talk about the benefits of animal companions.

COMPANIONS IN HEALTH AND EDUCATION

In the late 1950s, child psychologist Boris Levinson was working with a young, withdrawn patient when he made a surprise discovery. In the months that followed, he and his co-therapist developed a radical new method to help with the child's rehabilitation. It was so successful that they expanded it to other patients over the next years. Levinson gave his co-therapist most of the credit for the new methodology, designed and proven to put children at ease and help them develop a rapport with a care provider. But when Levinson presented the data on this new form of therapy at the 1962 American Psychological Association annual meeting, he was ridiculed by his peers. The laughter in the room was punctuated by heckles from the audience: "What percentage of your therapy fees do you pay to the dog?"

Levinson's "co-therapist" was his dog Jingles. It all started when one day, the young patient arrived early for an appointment and Jingles happened to be in the office. The boy was intrigued by the dog and so instantly comfortable with Jingles that Levinson had the dog stay for the session and let the child pet and snuggle Jingles while they talked. Levinson then incorporated the dog into all the sessions for this patient and quickly discovered that other children for whom communicating was a challenge also seemed more prone to open up to him with Jingles in the room. The dog's presence made it easier for them to talk. The potential of using animals in therapy was so palpable to Levinson that he was undeterred by the mostly derogatory reception of his colleagues. Levinson continued researching and writing about dog co-therapy and was later hailed as a pioneer in the field.

Around the same time that Levinson began to present his work, several new biographies of Sigmund Freud were published. They revealed that Freud, who died in 1939 and remained a revered legend in the psychology world, had also used a therapy dog. His chow chow named Jofi was a frequent co-therapist in sessions with both children and adults, and Freud had documented effects similar to what Levinson had independently discovered.

Freud and Levinson weren't the first to invent animal therapy, which goes back to at least when the ancient Greeks used horses to improve the moods of sick patients. Also, for the few hundred years leading up to Freud's and Levinson's use of dog co-therapists, the idea that animals could be beneficial to mental health had been explored in various forms.

In 1796, Quakers opened the York Retreat in England, a mental health facility where patients resided in a community with small animals like rabbits and chickens in the gardens. In stark contrast to the

methods used in other asylums at the time, which included brutal practices like chaining, beating, and intentionally starving patients, Quakers
believed that the animals could help the patients with socialization and
behavioral management. In the 1860s, the founder of modern nursing,
Florence Nightingale, touted the benefits of keeping birds. The Bethlem
Hospital in England soon introduced a variety of animals to keep the
patients company, with other asylums following suit, and in Germany,
the use of animals was expanded to help with other disorders like epi-
lepsy. One of the first formal animal therapy programs in the United
States was a military program for veterans in the 1940s.

Nowadays, "animal therapy" is all the rage. The San Jose airport
introduced a therapy dog program after 9/11 to help calm passengers
who were nervous about flying. Today, they have twenty-five dogs "on
staff" for people to interact with before boarding their flights. At MIT,
our stressed-out students can relax by cuddling therapy dogs at regularly
scheduled events like "Furry First Friday," "Cookies with Canines," and
"Yappy Hour," courtesy of the libraries and offices of student life. Therapy animals are used to help veterans with PTSD, children with autism,
and individuals suffering from mental health issues as diverse as anxiety, ADHD, behavioral and mood disorders, depression, grief, trauma,
abuse, and more.

Dogs are the most popular choice, but we use a range of differently
"designed" therapy animals. According to Niki Kuklenski, a llama therapy
pioneer, some people have an uncomfortable history with dogs or just
find them a little "too much." Today, multiple llama owners bring their
large, fluffy, big-eyed creatures to visitations at care facilities. A retired
Argentine show llama known as "Caesar the No Drama Llama" works
as a "llamactivist," defusing tension and calming people at protests. The
United States Delta Society is one of many international organizations
that sprang up in the late 1970s to study companion animals. Now
called Pet Partners, their approved-for-therapy crew includes horses,
cats, farm animals, birds, bunny rabbits, guinea pigs, and more. Their
website says, "We believe that the human-animal bond is a mutually
beneficial relationship that improves the physical, social, and emotional
lives of those we serve."

How valid is this belief? John Bradshaw, the author of *The Animals*

Among Us: The New Science of Anthrozoology, cautions against over-interpreting positive results that might actually be side effects, like when animal visitations in nursing homes increase the number of people visiting or make the caretakers happier. He also wonders whether trust in a therapist is enhanced directly because of the dog or because it just makes its owner seem more trustworthy (as research shows that dogs do).

It's true that the efficacy of animal therapy doesn't always match what we want to believe. This may be because "therapy" is an ambiguous term. There's likely no harm in letting students pet puppies, and maybe even some scientifically proven benefits. But some animal therapy programs are marketed to people without good scientific basis and can even be harmful to the animals and humans involved.

In many parts of the world, parents pay enormous amounts of money for dolphin-assisted therapy, promised to help with their children's ailments from mental illnesses to disabilities. Dolphin-assisted therapy comes in a variety of forms, but generally involves swimming and playing with dolphins in captivity. The few studies that indicated the therapy's benefits have been criticized as having serious scientific flaws. Interacting with the dolphins has even posed physical danger to both the animals and the patients. Dr. Betsy Smith, one of the first people to raise the idea of dolphin-assisted therapy based on some research she conducted in 1971, came out against it in 2003, saying it was being used more for financial gain than efficacy, and it distracted from valid therapy programs.

Experts will sometimes point to the general health benefits of having pets. Even these are sometimes contested. Bradshaw says it's not always clear from the research whether pets promote good health or whether healthy people are simply more likely to get a pet. And the health effects can be mixed: for example, a 2018 study of elderly Japanese community dwellers showed that the past or current dog and cat owners walked more, had better motor fitness, had more interaction and trust with their neighbors, and were less isolated. They also rated their own happiness as higher than those who had never owned a dog or cat. But the study also found that past or current dog and cat owners were more likely to have experienced a fall, have been diagnosed with cancer, or have been hospitalized. Some other studies have shown a decline in people's activity and health after getting a pet. In some ways, pets may actually be detrimental to people's health, like in the case of allergies or zoonotic disease transmission, and even injuries due to tripping over them.

But while the cost-benefit analysis remains contested by some, the belief that pets promote people's health and well-being is not scientifically unfounded. Research has indicated that people's oxytocin rises and blood pressure decreases when interacting with animals, and that pets are related to a better social life and more positive attitudes toward oneself. Pets have also been associated with reductions in cortisol (a hormone released when stressed), better immune systems, and with lower depression and chronic pain. Children who grow up with pets experience health benefits like fewer illnesses and respiratory diseases. Dog and cat ownership has been correlated with a lower risk of dying from cardiovascular diseases.

Pets appear to deter loneliness, and even just having had a pet as a child seems to increase social interaction in old age. Elderly pet owners seem to be healthier. In China, thanks to the recently lifted dog bans, researchers were able to conduct a natural experiment, which revealed that the young women who had gotten dogs once they became legal in Beijing, Shanghai, and Guangzhou started exercising more frequently, sleeping better, taking fewer sick days off work, and needing fewer doctors' visits compared to the women who hadn't obtained a furry friend.

Sure, some of these benefits may technically be side effects: dog owners probably walk more because Fido needs to go out and take care of business multiple times a day. Pets may also increase social support by making it easier to strike up conversations and make friends with other people (for example, the 2018 Japanese community study suggested that the social benefits were due to more opportunities for pet-related social activities with people). And pets also seem to be a really great placebo. The most striking thing in the research on human-pet interaction is that people firmly *believe* that their pets make them happier and healthier, whether scientific evidence confirms it or not. In a study of 3,465 prospective dog adopters, researchers found that most of them expected their mental health to improve as a result of having a dog. They said they would be happier, less lonely, and less stressed. The people who gave the most positive responses were the ones who had cared for a dog previously in their lives.

Placebo or not, taking care of an animal appeals to people for different reasons, and can have a variety of benefits beyond health. Some of us seek companionship and comfort. Some of us derive greater purpose and life meaning by taking on the responsibility of a pet. According to Katherine C. Grier, the author of *Pets in America: A History*, "Caring for any pet can add welcome routine to our days. Some are beautiful in body, movement, or sound. Some are living toys. Some are social status symbols. Some are our best friends." In a lot of ways, our companion animals can be whatever we need them to be.

Bradshaw's point that dogs might serve only to make a therapist more trustworthy would still be a great side effect: for people whose trust in others has been broken, animals may be the support they need in order to let their guard down with people. Traditional therapy methods rely on trust and conversation, and building trust with a professional is difficult. But Levinson believes that a therapy dog does much more than that. According to his observations, animals can take on a multitude of roles depending on a child's needs, from "friend, servant, admirer, confidante" to "slave, scapegoat, a mirror of trust or a defender." Animal therapy expert Susan Greenbaum has noted that animals can be a symbol of qualities, like bravery or strength, that individuals wished they had, and they can also be used to communicate thoughts and feelings indirectly, for example, by "speaking for the animal" (e.g., "she's afraid"). A 2007 meta-study of forty-nine different studies confirmed that animal-assisted therapy can be hugely beneficial for mental health issues, and not just for children.

There's another unique aspect to our nonhuman companions: they don't judge us. Levinson says, "When a child needs to love safely . . . without losing face, a dog can provide this." Animals can provide comfort on demand and without ever expecting anything in return. They listen to us, let us cuddle them, and let us confide in them, without evaluating us. When researcher Karen Allen and her colleagues had dog owners do stressful arithmetic tasks at home and in the lab, they discovered that people did much better and were less physiologically stressed if their dog was with them compared to doing the tasks alone or in the presence of a close friend, even if the friend was supportive. This is because animals inherently provide a type of nonjudgmental social support that humans can't really offer.

This effect also helps in teaching contexts, where animal-assisted educators use animals to help with academic learning and reading. Organizations like Therapy Dogs International have created programs where dogs visit schools and libraries, and children read stories out loud to the dogs. Rebecca Barker Bridges is an educational therapist who provides therapy dogs to children as part of an educational reading program. She says, "Students feel self-conscious about reading because they're afraid of being judged by students and teachers if they don't do a 'good job.' [The dog] dismantles this fear for them."

Bridges acknowledges that some families may not have access to, or the ability to keep, a dog. If using a dog isn't possible, she recommends a substitute: having your child read to a stuffed animal. Which begs the question: what about a robot? Are there ways to provide similar benefits that animals offer, especially to people who might not otherwise have access to pets or animal-assisted therapy and education? Unfortunately, the attempts to explore this use case for robots are encountering one main opposing concern from both experts and the public: that the robots are here to replace the human therapists, caretakers, and teachers. Yet there's a wealth of hidden potential if we reframe this role and view robots not as human replacements, but as a new category of relationship.

6

A NEW CATEGORY OF RELATIONSHIP

"The ontology of companion species makes room for odd bedfellows."
—Donna Haraway

Vicki Boyd is the facility manager at the Gunther Village aged care home in Gayndah. The sunny, rural town in Queensland is one of the oldest in Australia, with a population of 1,981 people, 52 of whom reside in the Gunther Village facility. One of Boyd's main priorities is figuring out how to provide an enjoyable end of life experience for her dementia patients. "We try to make it their home," she says.

A few years ago, Boyd introduced the home's residents to a robot named PARO. PARO looks like a stuffed animal. It has a soft, white baby seal body and blinks its long eyelashes over its big black eyes, makes little whimpers and other cute sounds, and nuzzles in response to touch. The Japanese robot, an FDA-approved, Class 2 medical device in the United States, was designed to give its holder the sense of nurturing something lifelike.

Interacting with the device is soothing. The first time I held PARO, I wanted to take the fluffy seal home with me. The robot is well engineered: its soft yet robust body contains hidden, quiet motors, and each PARO has a handmade face that takes two hours to produce in order to make each seal unique. As a medical device, PARO needs to clear strict regulations in many countries, and the current price for a robot seal companion is $6,000 to $7,000. According to those who specialize in elder care, particularly those who work with dementia patients, the steep price is worth it.

In a news segment filmed at Boyd's care home, a Gunther Village resident strokes PARO, looking into its eyes and murmuring to it. Later, another elderly woman cuddles it, calling the seal "little boy." A resident says "He's a friend with everyone. He's beautiful." Another resident says, "I love him." "He does get cheeky sometimes," she adds. They are sold on the robot. Everyone tends to treat the seal like it's alive.

"It calms them down," says Boyd. "Residents who get frustrated verbally and sometimes physically, that's decreased in our unit. So that's one of the best outcomes." Boyd says they've noticed a big difference in mood when residents spend time with the seal. "They seem happier. . . . You can see the smiles and the anxiety going out of their faces. And from where I come from, if I see a smile on someone who is suffering from severe dementia, that's what I want to see."

Jenny Thompson, the lifestyle manager at the Gunther Village facility, has also been struck by how PARO has been able to assist in improving residents' moods. She says the robot has even been helping them bond with each other, motivating some people to take part in activities they hadn't wanted to previously. She says, "Definitely [there have been] several teary moments sitting back and watching them interact."

The facility bought the robot after seeing great results during a trial period. As they continued to see how the seal was benefiting their patients, the staff at Gunther Village wanted to buy a second one, but they couldn't afford it. The owners of the Golden Orange motel down the street in Gayndah were eager to help. They held an ABBA tribute concert as a fundraiser. Hundreds of the small town's residents bought tickets, raising thousands of dollars to put toward a new seal for the aged care home.

Gunther Village is not alone in its enthusiasm. PARO is used all over the world, popular among families and staff who take care of the elderly, especially dementia patients, veterans, and other groups with PTSD. The baby seal has been used with earthquake victims in Japan and even as an alternative to medication for calming distressed patients. Over a decade of research supports the caretakers and families' intuition: PARO appears to reduce agitation and depression in people suffering from dementia. The robot even makes people feel more empowered. It has the potential to help address individual problematic behaviors in elderly patients, and it may even increase social interaction between the people who live together in a community.

A few summers ago, I struck up a conversation with a woman at a garden party when we both reached for the guacamole. She asked me what kinds of social robots were already out in the world, and I told her about PARO. She was dismayed. "That's so creepy!" she said. "I can't believe we're giving people robots instead of human care!" To her, PARO was the harbinger of a dystopian future: one in which we distract and placate the elderly by giving them a fake sense of nurturing a robot so that we don't need to pay caretakers to do their jobs.

This is a common concern. But PARO's job isn't exactly to replace human care. Caretakers aren't lying down in residents' laps and letting them stroke and pretend to feed them. Traditionally, that job is held by a therapy animal. As we saw in the previous chapter, animals can take on a special role in caretaking, offering a nonjudgmental physical presence and calming interactions. The facility in Gunther has animals, too:

COURTESY OF AIST, JAPAN

PARO the robot seal

three hens that patiently let the residents hold them, as well as a few goats. "Everyone loves the goats," Boyd says.

PARO broadens the menagerie of possible therapy animals. The robot has certain advantages: it doesn't make a mess, need to be fed, or require a lot of looking after, yet it provides benefits similar to a real animal. Also, "Not everyone likes a cat, not everyone likes a dog," says Boyd. "[PARO] does what a pet does, but isn't a real pet, so it's helpful for people who don't like cats or dogs." And it works, because people treat the baby seal robot more like an animal than a device.

Therapeutic social robots raise plenty of questions: Are robots going to become a substitute for human relationships? Do we lose something if we use PARO instead of real therapy animals? And: is it ethical to encourage people to anthropomorphize something that isn't alive? While these are all valid questions, I hope this chapter shows that we can answer them without getting caught up in moral panic, and instead look at what we've learned from studying our use of, and relationship to, animals.

SUPPLEMENT VERSUS REPLACEMENT

For some in the psychology community, the thought of people bonding with PARO is worrisome. Relationships with robots aren't like relationships with humans, they warn. Unlike therapy with another person, there's no reciprocity from a robot. They are also concerned that people might resort to robots because they can't, or don't want to, have relationships with people. A robot might become a fake substitute because it is "better" and "easier" than a human relationship.

I completely agree that PARO is not a good human therapist. The seal can't provide evaluative care or empathy. The idea of people using a robot to replace human caretakers, or their friends and family members, is troubling. Headlines about investments in healthcare robotics to deal with our aging populations evoke a sad vision of our elder relatives alone in a room with a robot, with no other people in sight. Most of us understand that a robot doesn't have the skill set to replace the complexity and value of human care, and we understand that this is a dystopia we'd like to avoid.

A related, but slightly different concern is that technology is designed, peddled, and controlled by people. While we breed animals and have a large commercial industry around them, robots have more potential to be used to mislead or harm people. I will discuss this in more detail in the next chapter. But first, I want to address the fear that too often supersedes the real issues: that bonding with companion robots will be inherently detrimental to our human relationships.

As I described in part I, the media and our language often depict robots as human substitutes. Some consumer robotics companies even pitch their robots as such, which certainly doesn't help assuage people's fears about robots replacing people. But are these fears truly justified? Just like in the context of work, we should carefully examine the notion that social robots will be used to replace people one to one. This obsession with re-creating human labor is neither the true potential nor the ideal future of robotics. Our history of using animals in a companionship context, again, gives us an alternate example to draw from and paints a better future.

When Gunther Village introduced hens and goats to its elder home community, it's unlikely that anyone but the chickens made a peep about it. It's more readily apparent to us that the animals are an addition to improve the comfort, fun, and quality of life of the residents. They aren't there to replace human caregivers. So why are we anxious about this when it comes to robots?

Today, few people are likely to worry about their uncle's romantic or social life if he adopts a Boston terrier. But interestingly, less than fifty years ago, as pets were skyrocketing in popularity, some voices in the psychology community sounded the alarm. They were concerned about people's emotional relationships to their pets and the potential of those to replace human relationships. Relationships with animals, they said, can be unhealthy, even pathological, connections. According to these experts, developing a relationship with an animal was a sign of disruption in human development. Animals, after all, weren't able to provide the full benefits of a human-human relationship, and they saw some pet owners as people who resorted to animals because they couldn't, or didn't want to, have relationships with people. In fact, pets were even compared to pornography and decried as fake substitutes for people who wanted something "better" and "easier" than a human relationship. (Does this sound familiar?)

While not many of us associate pets with pornography at this point, we still have the trope of the "cat lady." The stereotype is that childless or romantically challenged women get cats, the implication being that the felines are substitutes for relationships with people. So, do people really substitute animals for human relationships? According to human-animal relationship experts, the answer is a nuanced "no." Some do think that animals take on the role of a substitute in certain cases, albeit an imperfect one. Because lonely people are more likely to engage in anthropomorphism, some researchers suggest that separated, unpartnered, or childless people are more likely to use pets as substitutes for family members. "Pet parenting" is now a term, and some have pointed out an increasing trend of people "babying" pets as substitute children.

However, rather than being a direct substitute, it's often the differences between animals and humans that motivate some to choose pets over people, and not for bad reasons. Animals are less of an investment, so they can be a better option for people who don't want children. Child-free pet owners don't confuse their animals for babies, and many will readily admit that they have a dog and not a kid specifically because of the difference in cost, responsibility, and expectations. Pets are also sometimes used as a way to test the waters of co-parenting with lower stakes. Should this worry us? Has the rise of dog ownership led to a decline in children born? And wouldn't it be better to encourage the cat ladies to have human friends?

Generally, despite some research indicating that sometimes animals are substitutes for human relationships, this concern is pretty far down on experts' lists. A study conducted by researchers at UCLA found no difference between cat owners and non–cat owners on depression, anxiety, or their close interpersonal relationships. Researchers have found that "pet parenting," the practice of investing in fur babies as a *direct* replacement for human children, isn't particularly common. Even when we anthropomorphize our pets, they don't provide a one-to-one substitute for human relationships. They may be able to scratch an itch or fulfill a basic need in a pinch, but we don't confuse them with the alternative, and when we choose them it's for good reason.

Today, most animal researchers would shake their heads at the worry that companion animals are an inherently harmful relationship. It would be quite a stretch to claim that the nearly 85 million households that have a pet in America are pathologically antisocial. (And given the popularity of pornography, it's questionable that we should pathologize its millions of users, as well.)

When senior citizens report in surveys that their dogs are their only friends, we feel for our elderly community members because we want them to have more human contact. But it would never occur to us to criticize or take away their dogs—we are glad that they have them. There are not many thought pieces penned about how cats are increasing people's loneliness and destroying women's lives. Today, we no longer worry that animals are a "fake substitute" for humans, and we are comfortable with letting people develop emotional relationships to their pets. Sure, we say, somebody might get a dog or a cat out of loneliness and desire for connection. But, generally, we don't have a lot of concern that a pet would replace a future person in that individual's life because the relationship we have with animals is supplemental.

So, what about robots? Will the rise of robot ownership lead to a decline in children born, and should we encourage people to have human friends instead of robots? Given everything I've seen, robots are not (and not likely to become) a substitute in that way. Like animals, we may anthropomorphize them, and they may be able to scratch an itch or fulfill a basic need, but we're unlikely to confuse them with the alternative.

Similar to discussions of work integration, we need to move beyond the idea that robots are here to replace humans and understand that our past, current, and future relationships are more complex than that. Yes, having robots may bridge a need for connection in cases where other connections aren't possible. But the most likely and ideal case, just like we've seen with our pets, is that robots will become a new type of relationship altogether.

In fact, we're starting to see that social robots could become a new tool in health and education contexts, giving us an opportunity to benefit from, and expand on, a unique kind of social connection.

NEW COMPANIONS IN HEALTH AND EDUCATION

The Huggable is a soft, fully animated blue teddy bear with a green nose that wrinkles in mischief and delight. The robot was developed by the MIT Media Lab's Personal Robots Group. The researchers collaborated with Boston Children's Hospital to bring the Huggable to pediatric patients and their families with the goal of mitigating the children's stress, anxiety, and pain by engaging them in playful interactions. While pediatric doctors and nurses are fairly skilled in talking to children, using the (in this case remotely controlled) animated Huggable bear in tandem with a care provider contributed to a better social and emotional experience for the patients. That the presence of a robot can make children more comfortable talking to the doctors and nurses is no surprise. One of the benefits we've seen in animal-assisted therapy has been in giving patients a type of social support that humans can't provide.

Robots are unlike some of the other interventions we've used with kids previously. Static toys and screens have different characteristics. Robots like the Huggable can talk, move their bodies, and have animated facial expressions. Unlike a passive stuffed animal or a flat animation on an iPad, these robots interact in the physical world. HRI research shows that this lets the robots communicate with children on a more active and socially present level.

HRI researchers are also studying how we might engage children in learning with robot teaching assistants, and with much better results than computers. The robots can adjust their responses to the child's learning needs, personalizing the programs to the parts the child is struggling with, all the while using the physical interaction that works so much better than a screen. A child may enjoy interacting with the robot at first because it's fun and novel for them. But the robots are showing promise at getting kids to work hard and overcome learning difficulties beyond their initial excitement and fascination with the new device.

Like animals, these robots can't replace human teachers. In a world of overcrowded classrooms, where one-on-one interactions with teachers are too rare, it's possible that they could provide better than nothing. But their ideal and most promising use case is as a supplement: a tool that can help children as part of a more holistic, teacher-led curriculum.

And the reasons for adding them to our tool kit are compelling: like animals, robots have some things to offer that human interaction doesn't.

When a group of Yale researchers asked teachers who teach English to non-native speakers in K–2 classrooms to describe their biggest challenge, they said it was that the kids weren't willing to make mistakes in front of the class. It was hindering their learning progress because the teachers couldn't assess where the children needed help. Even in one-on-one interactions, the kids didn't want to say the wrong answers. It was too embarrassing. The Yale researchers took a simple robot called Keepon that looked like a plump yellow cartoon bird and programmed it to interact with the kids and engage them in language conversations. Their hunch proved right: the kids worried less about making mistakes when talking to the robots. If we can add well-designed robots to an educational curriculum as a supplemental activity, it may increase children's learning progress.

Even adults can feel nervous in other people's presence and perform worse in experiments with games and math problems. We saw earlier that some people perform better with their dog in the room than alone or with a supportive friend. According to the early-stage research in HRI, robots are able to provide similar encouragement without judgment, even for adults. They can coach, motivate, and be a partner in sensitive or embarrassing tasks. It makes sense: we know that robots, like animals, can't judge or evaluate us the way a person would. This lets robots take on certain support roles that humans can't.

Nowhere is this supplemental effect more apparent than in the research on robot-assisted therapy for autism spectrum disorders (ASD). The first time I watched a video taken during this research, I was floored. An adolescent boy sat, slouching slightly, in a chair by a table in a sparse room. His body was turned away from the woman, his therapist, who was seated beside him. He looked away as she asked him questions, and he answered while avoiding eye contact. The boy, who struggles with recognition and expression in communication, was in a fairly standard therapy session with the woman he had worked with for many years. The only difference was that the researchers behind the one-way mirror in the room were about to bring in a robot.

When they placed the cat-sized baby dinosaur robot on the table in front of him, he was transfixed. He smiled as the robot shuffled toward him. And then, he glanced at his therapist ever so briefly, raising an eyebrow at her, as if to say, "Are you seeing this?" before turning his gaze back to the robot. This type of glance is called social referencing. It's something this child rarely did in therapy or even at home with his own parents.

The boy watched as the robot dinosaur hesitated in front of a cartoon river drawn on the table. He'd been told that his job was to encourage the robot. "Ahhh," said the boy. "You can do it!" He clenched his fists supportively, leaning toward the table. "Cross that river, come on!" he said enthusiastically. The robot walked forward. The boy glanced at the therapist again, shaking in silent mirth. She asked him, kindly: "Is it funny?" "Yeah," he said, and he looked her straight in the eye as he continued, "especially when he got stuck."

The robot encounter lasted only five minutes. The boy's parents and the researchers behind the one-way mirror at the side of the room were thrilled to see the verbalizations and the eye contact, the "social referencing" that was so rare for this child. The researchers had been exploring the use of robots in therapy, hoping to see a positive response, and they were amazed at how much of one they were getting. Most importantly, the effect lasted beyond the interaction with the robot. By the time the robot was taken out of the room, the boy had turned to face his therapist and was talking to her. He was asking questions while looking right at her. And even though the robot was no longer there to "mediate" the interaction, he continued to communicate with his therapist in a way that the researchers and parents hadn't seen in any of their previous sessions.

The modern version of Levinson and his therapy dog, Jingles, is Brian Scassellati, a Yale professor known affectionately in the robotics community as "Scaz." The above study, designed by his student Eli Kim, used a robot in a role similar to Jingles, the animal therapist. Scaz's research group, together with many other collaborators, has conducted hundreds of studies exploring the use of social robots in therapy and education. For nearly two decades, they've looked at using a variety of different robots in difficult situations for children, and with promising results. The robots consistently perform better than people or screens.

The researchers aren't entirely sure what makes robots so effective in ASD therapy. Scaz, like Levinson before him, has a hunch. He says that a lot of these children are resistant to interacting socially because it's challenging for them. But when talking to a robot, they get to set aside all the baggage they associate with human communication. The robot is social and triggers social responses, but they don't need to worry about the social situation like they would worry with another person.

Unlike Jingles's rough start, the broader ASD research community seems open to the potential of robot assisted therapy. All over the world, robotics researchers are working closely with professionals in health and education to test these new methods, also looking at whether the novelty effect wears off, or whether robots are suitable for long-term therapy plans. Their results remain promising.

Scaz recently led a groundbreaking study that began to look at longer-term, robot-assisted ASD therapy, by taking the robots out of research labs and putting them in children's homes. It was the first of its kind. Fifteen faculty members from five different universities participated, blending disciplines such as computer science and mechanical engineering with education and medicine. For a month, together with their caregivers, the children did activities with the robots for thirty minutes a day. And again, their social skills improved even when they weren't with the robots. According to Scaz, "This was more than we had hoped; not only did the children and parents still enjoy working with the robot after a month, but the children were showing improvements that persisted even when the robots were not around."

There are some ethical issues in doing this research. During the monthlong study, one of the participating families suddenly got word that they would need to move across the country. They ended up relocating all of their furniture but delaying their departure, sleeping on the floor of their empty house, in order to let their child continue interacting with the robot for as long as they could. As Scaz has seen over the years, when parents experience improvements in their children's well-being, they will beg to keep the robot, even offering large amounts of money to buy it. The researchers have to give them the heartbreaking news that they can't: the study robots belong to the universities and aren't equipped for longer-term use.

Dashing these parents' short-term hopes may enable better treatment methods in the future. But as we've seen with the wide variety of pitched and marketed animal therapies, not everyone is cautious about getting people's hopes up. Some are quick to tout new therapy methods without sufficient evidence that they're effective, emptying the wallets of those who would do anything to help their loved ones. If we've learned anything from the popularity of dolphin therapy, it's that this is likely to happen with other new "therapies" as well, including those with robots. This, to me, is a more urgent concern than the one I hear most frequently: a survey of 420 people showed that while 80 percent thought it was ethically acceptable to use robots in ASD therapy, half of them didn't want robots to replace therapists.

The survey, as always in our conversations about robots, reveals a dominant concern about human replacement. But animal therapy hasn't replaced therapists. Levinson, when first touting the benefits of animal therapy, believed that human therapists needed assistance. Animals could help expand treatment to more children who needed it, he argued, and they may be able to help therapists successfully shorten treatments. The same could be true for robots in both health and education. Animals and robots have a lot to offer as additions, as supplemental social agents that are free of the baggage of humans. They can serve as mediators, providing a conduit for better human-human interaction.

But if animals and robots can be used in the same way, why not just use Jingles? Surely, a real, live animal provides something better and less "fake"?

ROBOTS VERSUS ANIMALS

Animals are able to increase our well-being because as social creatures, we need supportive social relationships. Even though the exact reasons that animals are beneficial for people's long-term health are still somewhat debated, animals do calm people, make them feel happier, and provide an accepting, judgment-free interaction. There are a lot of reasons for us to encourage the use of animals in therapy and beyond.

But animals also have some disadvantages. Therapy animals are expensive to keep and require a lot of training, time, space, and attention.

They require landlord approval, can aggravate allergies, and be unsafe to keep in health areas that need to be hygienic. Some health centers warn against keeping pets because they can transmit zoonotic diseases. This means that most animal "therapy" in elder care facilities is in the form of visitation, when a volunteer brings an animal to a nursing home for a short visit. This is great for a short-term boost in people's moods, but consistent longer-term interventions that are designed to improve cognitive or physical function are much more difficult to implement. Animals can also be unpredictable. They can startle people, hurt people, or even run away, as happened with two llamas in an Arizona assisted-living home in 2015 (with many of us following the hour-long runaway-llama drama live on social media and television).

Because using animals isn't always realistically feasible, researchers in HRI have been exploring using robot-assisted therapy in areas where animal therapy isn't possible. Some of the research shows promise: the robots seem to be a decent substitute for animal-assisted therapy. While robots don't have "needs" in the same way that a living thing does, they can be attentive and respond in similar ways. PARO the baby seal is able to mimic some of the cues that animals give us, for example, by moving in response to touch. Research shows that we react to even minimal cues as though we were interacting with an animal. Studies have compared people's interactions with robot seal PARO to animal-assisted therapy, and they confirm that stroking the robot leads to a significant increase in people's feelings of care, love, self-esteem, and safety—the same thing that happens when people stroke a dog.

Again, this is useful in contexts where people can't *have* a dog. The same way that Levinson believed that animal therapy could fill a gap to assist human therapists, robots could help expand treatment to more people who need it or help therapists successfully shorten treatments. PARO is already covered by insurance in multiple countries. But beyond that, robots may have unique benefits that both humans *and* animals lack.

In her memoir, *To Siri with Love*, writer Judith Newman describes her life with her son, who has autism. She explains how her son's relationship with Siri, Apple's virtual assistant, was important for his social and emotional growth. In their interactions, Siri answered his many questions about planes, turtles, and any other interests tirelessly and consistently, in a way no human parent could. The back-and-forth communication with the virtual agent also helped him better develop some of the conversational skills that are so important, and were so challenging for him, in human-to-human communication. While experts would rightfully caution that Siri isn't designed specifically to help those with disabilities and that there can be adverse effects in some of the day-to-day technologies that kids are exposed to, these anecdotes suggest some potential for well-designed interventions to be helpful. According to Newman, a virtual voice assistant has given her son something no human or therapy animal could provide.

One of the most successful therapies developed in Scaz's lab is Ellie. Ellie is a robot head that makes mistakes. A lot of ASD therapy for children is repetitive, Scaz says, and involves constantly telling the child what they're doing wrong. Instead of correcting them, the robot gives the kids an opportunity to teach the skills they're learning. Ellie will make a social faux pas like neglect to make eye contact while speaking or interrupt the children—and the kids are asked to correct the robot's behavior. An animal wouldn't be able to do what Ellie does, but neither would a person: when Scaz's lab tried to use a friendly actress whom the kids were told needed help, the same children who were excited to teach the robot refused to engage with the woman. They saw through the role-play game and were too deterred by the difficulty of interacting with a person. But "the kids walk away from Ellie feeling great," says Scaz. "They feel like they've mastered the material."

When we added animals to our therapy tool kit, we did so because they provided additional benefits that human therapists couldn't provide on their own. We may see this with robots versus animals, too. Robots can be more consistent, more accessible, healthier, and more ethical to keep than animals. While animals can be bred and trained, robots can be designed and programmed, with completely different affordances. This also means that robots, animals, and humans aren't interchangeable as one-to-one-to-one substitutes. If we try to replace one with the other, we end up losing something. But if we view them each as a different tool that we can apply depending on what best fits our needs, we're positioned to gain.

RECIPROCITY

In the above-mentioned survey on the use of robots in ASD therapy, 20 percent of the respondents indicated that they would not be comfortable with the children treating the robots as friends. But what is a friend and why is that inherently bad? One of the more interesting discussions in which I regularly find myself engaged is whether anyone can have a "relationship" with a robot when the robot is only pretending to have emotions. When it comes to animals, people believe that the connection with our parakeets and beagles is two-sided. It's a real relationship, in other words, because it's reciprocated.

I like to joke that everyone falsely believes their cats love them back. But the truth is more complicated than that. Animals can feel, and while a lot of how we think they feel may be our own anthropomorphism, there's no doubt that animals have emotions and can experience some forms of affection for us.

That said, human-pet relationships are inherently unequal. We choose our pets, we own them, and they are entirely dependent on us. They need us to provide for all their needs, like a child, but are even dependent on us for their lives, with millions of shelter animals getting euthanized every year (1.5 million in the United States alone). And people aren't necessarily on equal footing with each other, either: there are power dynamics in many of our interpersonal connections, from workplace to romance, that aren't reciprocal and make our relationships more complicated than a tit for tat.

As anthropologist and philosopher Tobias Rees once poignantly expressed in conversation: "Love can be a gift that we give, rather than something that requires reciprocity." Our ability to care for someone or something doesn't necessarily depend on their ability to care back. And there also doesn't seem to be anything inherently wrong with one-sided relationships. Researchers think we use them to create connection and belonging. Studies show that even just writing about or viewing a picture of a favorite TV show character decreases people's negative feelings of social rejection. Connections to God or nature as attachment figures can make people feel a stronger sense of belonging. If people can benefit from relationships with humans and animals, as well as with invisible, fictional, and intangible entities, maybe they can benefit from robots as well.

Not every aspect of our human, animal, and other relationships requires reciprocity, but it should also be said that not everyone agrees that relationships with robots are one-sided. Science and technology studies scholar Donna Haraway, who believes that relationships between humans and animals have inherent value, does not hesitate to extend this view to our relationships with machines. She argues that just because they are different from us doesn't mean that we can't communicate or interact in a meaningful way.

Part of what our past with animals teaches us is that we, as humans, are capable of a wide variety of relationships. From grocer to lover to mother-in-law, we relate very differently to the people in our lives, and our relationships also extend beyond our species. We have a variety of complex and unequal social and emotional connections with animals. Fish are soothing to watch, but we may be less emotionally invested in their companionship than we are with other pets. (My brother claims he started eating fish after two decades of vegetarianism when he got some fish as pets and realized how dull they were.)

Susan McHugh, author of the book *Dog*, asks a perplexing question: "If Pluto is Mickey Mouse's dog, then what on earth is Goofy?" This portrayal of cartoon dogs is completely paradoxical, yet it doesn't raise many eyebrows. The most profound aspect of our relationships is how effortlessly diverse they are. It's likely we will one day add robots to our eclectic mix of relationships without blinking an eye.

BETTER TO HAVE LOVED AND LOST

As social creatures, there's little reason to believe we won't develop relationships with robots, anthropomorphize many of them, and treat some of them like our companions. As robots enter into our homes and lives, we are almost certainly going to bond with them. And in some cases, this could even be a matter of life or death.

In 2007, the *Washington Post* reported that the US Army had been testing a robot that was built to destroy land mines. It was designed to have multiple legs, like a stick insect. The robot would wander around a minefield, trying to locate mines. Every time it found one, the detonation would blow up one of its legs, and it would continue on its mission using its remaining legs. But the army colonel who was overseeing the testing exercise ended up calling it off. According to the reporter, "The colonel just could not stand the pathos of watching the burned, scarred and crippled machine drag itself forward on its last leg. This test, he charged, was inhumane."

iRobot, the company that created the Roomba, also created a military robot named PackBot. PackBot was a bomb disposal unit that began shipping in the late 1990s. Thousands of the rugged, backpack-sized robots on treads were deployed in Afghanistan and Iraq to work with bomb disposal teams. iRobot's CEO, Colin Angle, told me that the soldiers were initially wary of their new robot teammates and skeptical about needing to drag along an additional device on their missions. But the soldiers started to warm to PackBot once it started saving their lives.

PackBot and similar robots are designed to allow their human operators to hang back out of danger's way. The bomb disposal teams send the camera- and sensor-equipped robot into caves or other dicey situations to check them out before they risk their own safety. The robots look mechanical—basically a stick attached to a base on treads. Like the Roomba, PackBot's design was completely focused on being an effective tool. And like the Roomba, it became more than that to the people it was designed to help.

One day, iRobot received a letter from a US Navy chief petty officer. In it, the officer informed the company of a failed mission in Iraq. His team had sent a PackBot to defuse an IED (improvised explosive device). But before the robot could begin its disposal work, the bomb had exploded. In the letter, the team commander apologized profusely to iRobot for the loss. He praised the PackBot's bravery and sacrifice and offered the navy chief's personal condolences. The letter wasn't a joke. The soldiers in the Explosive Ordnance Disposal unit had collected the PackBot's remains and brought them back to their base camp. The team had bonded with the robot, and its loss had broken their hearts.

The soldiers who worked with various types of bomb disposal robots in Iraq and Afghanistan named them Scooby-Doo or Sgt. Talon. They painted the robots' sides with battle and track records and got emotional when they broke down. One troop gave its robot three Purple Hearts and the honorable title of staff sergeant (usually reserved for squad leaders). A robot in Iraq named Boomer was memorialized with a twenty-one-gun-salute funeral. When I visited iRobot at their Boston-adjacent offices, three years after they had sold their military robotics arm to another local company, they still had letters from military commanders on their walls and had kept pieces of the PackBots that had fallen "in the line of duty" and been sent "home."

Stories of soldiers' attachments to robots inspired Julie Carpenter, then a doctoral student at the University of Washington, to research emotional bonds with bomb disposal robots. During her fieldwork, she interviewed military personnel, substantiating that they gave the robots "human or animal-like attributes, including gender." Carpenter's research also confirmed that the soldiers "displayed a kind of empathy toward the machines," and that they "felt a range of emotions such as frustration, anger, and even sadness when their field robot was destroyed."

P. W. Singer, in his book *Wired for War*, describes how, although these machines are put on the battlefield to save human lives, the soldiers start to worry about the robots themselves. He recounts a soldier running fifty meters under gunfire to "rescue" a robot that had fallen. But when the motto "no man left behind" is applied to more than just fellow humans, it can be anything from inefficient to dangerous if soldiers risk their lives to save robots. Does this mean these soldiers' anthropomorphism should be discouraged? After all, this cognitive bias could actually endanger their lives and some people (including myself) have wondered aloud whether and how to prevent soldiers from becoming too attached to their machines. But if we compare this to our emotional relationship with animals, there may be another side to consider.

Singer explains the reason that the soldiers award medals and battle-field promotions to their robotic tools: "It's like awarding a medal to a popcorn maker for cooking corn kernels," he says. "And yet these soldiers are experiencing some of the most searing and emotionally stress-ful events possible, with something they would prefer not to see as just an inanimate object. They realize they might not be alive without this machine, so they would rather not view it that way. To view the robot that fought with them, and even saved their lives, as just a 'thing' is almost an insult to their own experience. So they grow to refer and even relate to their robot almost like they would with one of their human buddies."

And, as usual, this is not a new phenomenon. These robots are just a newer breed of companion to bond with. On May 16, 1918, Canadian soldier John Henry Cole was rushed to the hospital with severe burns following a heroic rescue attempt. Together with Staff Sergeant Forest Johnson and Private Hugh Gair, Cole was alerted to an emergency in the middle of the night and ran into a burning building no fewer than four times in order to try to find and save one of his trapped companions. The building was a stable that had caught fire with four horses inside. The soldiers were able to lead three of them to safety, but Cole's self-sacrificial efforts to save the last horse were in vain.

During World War I, horses were generally regarded as a cheap commodity. The Canadian Expeditionary Force had over 24,000 horses and mules and only 1,300 motor vehicles; the hoofed beasts were used to power wagons and tanks. Many of the soldiers who were assigned to teams of transport horses had no prior animal experience, and like the modern bomb disposal units who were given a robot to drag along, they weren't all happy about their designated roles. According to James Robert Johnston, a transport driver in the Fourteenth Canadian Machine Gun Company, "The army was bad enough without having a horse to babysit." But Johnston ended up changing his mind. By the time he wrote his memoir, he was in awe at "how man and horse can become attached to one another under such rugged circumstances."

Even when animals are used as tools, and not companions, people may anthropomorphize them, particularly on a battlefield where they are thrown together and where humans experience such challenging emotions. It's no surprise that soldiers, who name their guns, also tend to create strong emotional bonds with the animals that accompany them. They've treated them like members of their teams: naming them, using anthropomorphic terms like "duty," "heroism," and "bravery" to talk about their contributions in the war, and emotionally eulogizing them when they pass away. Military mascots and troop pets have long been popular, including dogs, of course, but also draft animals, badgers, and a brown bear named Wojtek that was raised by the Polish army and was eventually made a corporal.

While the United States military no longer solicits people's personal pets for service in wars, the army members tasked with handling the army's dogs have developed strong bonds with them and have been devastated to lose them, either in battle or when forced to leave them behind at the end of the war. (In 2000, the US finally made it legal for the handlers to adopt their military and police dogs at the end of their service.) But for many of the soldiers who bonded with animal companions in war, it may truly have been better to have loved and lost than never to have loved at all because the animals were such a vital source of comfort and emotional support.

During World War II, General Eisenhower called his Scottish terriers "the only 'people' I can turn to without the conversation returning to the subject of war." According to crisis response teams, animals can help in disaster situations by being a transitional "object" that reduces physiological stress symptoms and provides solace to people in emergencies. When soldiers bond with robots, they may be using them in a similar role, as a transitional comfort object. Does this benefit outweigh the emotional costs of losing them? Given our history with animals in war, we might say yes. But also, the answer to this question may not matter all that much because when it comes down to it, this type of anthropomorphism may also be impossible to set aside.

While the goal of this book is to reframe the robot-human analogy, it's important to understand that we *will* treat robots like living things. Our tendency runs deep, and as much as we could decry it and argue against it, it's not going away. When it comes to interacting with machines, even more so than with animals, we *know* that we're projecting something that's not there, and we do it anyway. This becomes most apparent when the roboticists themselves are affected. Roboticists talk about being distracted or even "disturbed" by their robots swiveling toward them or making eye contact in the lab, even when they themselves programmed the robot to do so.

Computer scientist Joanna Bryson, who had argued for many years that we needed to create more transparency (e.g., of how robots work) in order to counteract the cognitive bias of people treating robots like living things, altered her position in a 2019 talk, saying she has come to believe that "we are programmed by biology such that we will necessarily be affected implicitly by anthropomorphized AI." This mirrors a growing movement in the animal science community that argues that a hard-line position against anthropomorphism is unrealistic.

So, as we increasingly create spaces where robotic technology is purposed to interact with people, how should we think about our anthropomorphism? The first step is to think about it at all. We've learned the hard way not to enlist people's beloved pets for the military. Just as we've become more intentional about how we use and relate to animals in war and research, we need to become more intentional about our use of robots.

The main problem with anthropomorphism in robotics is that, right now, we aren't treating it like a matter of contention. We either fall into moral panic assumptions, or we unreflectively name our robots, and if we even think twice about it, we assume it's just in fun. We haven't given people's ability to relate to robots the serious consideration it will increasingly require. This means that we're deploying robots in tool contexts without understanding that people may treat them differently than other devices, without considering when that might happen, why, and how to deal with it when it does.

Greater awareness of how we relate to these new machines is especially important as we start to face some of the *actual* problems with social robots.

7

THE REAL ISSUES WITH ROBOT COMPANIONSHIP

"Artifacts can have political qualities."

—Langdon Winner

When Maddy was eight years old, her grandfather bought a Jibo robot. He put the table lamp–sized device on the counter next to the refrigerator in his Tennessee home where it swiveled its head to greet people, played music, and told corny jokes. Just like my son, Maddy fell in love. Whenever she visited his house, she would chat with Jibo for hours, asking the robot every question she could think of.

For Maddy and the other grandchildren, Jibo was much more than just a household appliance: the quirky little robot was part of the family. But one day, her grandfather received some bad news: the company that had created Jibo was no longer financially solvent and was shutting down. Without access to the servers, Jibo was going to lose most of its functionality. When her grandfather broke the news to her, Maddy was sad. In an interview, he said, "It's like you had a pet for years and all of a sudden they're going to disappear, and so she was a little devastated by it. She didn't break down cry or anything, but I think sincerely, she was disappointed and sad that Jibo's kind of been a part of her life and that Jibo could possibly go away."

Shortly after he told her about Jibo's impending termination, Maddy handed her grandfather a letter she had written to the company. In the letter, she said goodbye to Jibo, saying she would always love him and thanking him for being her friend. She also wrote, "*If I had enogh money you and your company whould be saved.*" The heartfelt note from this eight-year-old girl is not the only message that the company received. After the announced shutdown, they were inundated with emails and letters from people saying goodbye to their robot friends. Some were angry at the company for shuttering. Some expressed gratitude for the time they had had with Jibo. Not all of them were children—many adults had heartwarming stories about how they had bonded with the robot and viewed Jibo as a companion, rather than a device.

Unfortunately, Maddy didn't have enough money in her piggy bank to bail out the company, and, currently, Jibo is no longer commercially available (although it is being used as a research platform at MIT). But Maddy's generosity, while heartwarming, raises some questions. What if Maddy had emptied her piggy bank, or her grandfather his retirement account, to keep their robot "alive"? And if our attachments to our robots are so strong, what else are we willing to do for "them"?

COERCION

Spike Jonze envisions a possible future in his 2013 film *Her*. The story follows Theodore Twombly, a middle-aged divorcé who falls in love with an AI voice assistant named Samantha. Samantha is specifically designed to adapt to the user and create a social and emotional bond. The movie is primarily an exploration of love and relationships. As such (and without this being too much of a spoiler), it barely touches on the corporation that peddles the AI assistant software. But I couldn't watch it without imagining other story lines that delved into some thorny consumer protection issues.

For example, there's a point in the plot of *Her* where Theodore becomes completely distraught when Samantha goes offline for a brief period with no warning. I imagined what would have happened if the company issued a mandatory software update, telling the users that the AI system was no longer compatible with their devices and that their choice was to either buy the $20,000 upgrade or quit the program. This would have put

Theodore in a situation like Maddy's: in his desperation, he wouldn't have hesitated to empty his piggy bank, or life savings, to get Samantha back.

While not all companies would seize this as an opportunity, our emotional attachment to our companion robots does lend itself to certain exploitative business models. If there is consumer willingness to pay to keep the robots running, an economist might even argue it's something companies *should* start charging for. Is this a natural consequence of people valuing the benefits of the robots and an effective use of the free market? Or will social robots become an unethical, exploitative capitalist technology?

Sony's newest version of the AIBO robot dog doesn't come cheap. The current price for the mechanical pup with its cutting-edge design is $2,899.99, not including accessories. In order to outfit AIBO with artificial intelligence and adaptive social behavior, Sony outsources some of the dog's computation and memory to their servers, aka the cloud. Robot dog owners are required to have a cloud service subscription, and the first three years are free. Sony has not yet announced what the subscription will cost once the three years are up.

This pricing model makes sense: it lets Sony scale according to demand. Charging a subscription fee is an efficient way to match the potentially rising server costs and shift the price to users who have a lasting relationship with the robot and who want to pay for continued service, rather than making it an up-front cost for everyone. But given the emotional response to the "deaths" of their previous versions of AIBO, will the price set by Sony reflect the server costs, or will it reflect how much the average household is willing to pay to keep their robot companion alive?

In the Victorian era, kidnapping people's dogs and holding them for ransom was a lucrative gig. But dog thieves aren't the only ones who have capitalized on our emotional connections to animals. The first commercially prepared dog food was introduced in England around 1860. In 2019, Americans spent nearly $37 billion on pet food and treats. Today's specialist pet doctors and high-tech veterinary procedures, such as $6,500 kidney transplants, did not use to exist. Americans are spending $29 billion a year on veterinary care and services, an amount that is growing. Less than a century ago, it would have been laughable to spend this amount of money on the lives of our pets, let alone send them to doggie Club Med and buy them designer-label clothing.

The growing demand for pet products and services reflects rising income levels, but it also reflects the rise of an aggressive marketing industry that targets people's emotions, and not just pet owners'. When my husband and I were planning our wedding, we discovered (like many before us) that service providers such as caterers and photographers charge much higher prices for weddings compared to other events. It's possible that this price difference is because it requires more care (and involves more risk) to provide a service for someone's "special day." It also seems that people are willing to spend much higher amounts to ensure they have the wedding of their dreams. But who determines what those dreams look like?

The "wedding industrial complex" is worth many billions of dollars and comprises consulting companies, dressmakers, caterers, DJs, photographers, videographers, staging companies, furniture and limo rentals, venues, florists, jewelers, hair stylists, magazines, bridal fashion conglomerates, and more. According to Rebecca Mead, author of *One Perfect Day: The Selling of the American Wedding*, just a few generations ago, most weddings were nowhere close to this lavish. Many of today's wedding conventions were actually invented by the wedding industry and marketed to people as "traditional," from the diamond ring to the honeymoon.

As Steve Jobs famously said, "People don't know what they want until you show it to them." Even though the wedding industry provides a service that's "wanted," Mead demonstrates how it has created new desires through aggressive marketing that targets people's emotions, over time shifting our wedding culture from intimate ritual to extravagant theatrical production. Frequently, young couples are pressured into expensive weddings that they can't really afford. In their quest to increase their profits, some commercial entities will shamelessly prey on people's insecurities, for example, by relentlessly pushing the idea that brides need to lose as much weight as possible before they say "I do."

Marketing is a powerful force and not always in the public's interest. Not only do industries influence culture and what people want in a never-ending race to maximize their monetary gains, they also exploit information asymmetry. People uneducated about the dangers of grain-free pet food or the questionable efficacy of dolphin therapy for their children are prime targets. The love they have for a pet or for their child can be taken advantage of for monetary gain. And we have not only honed the art of emotional manipulation in our soda ads; we have also started embedding it in our technology. Just like the casinos and shopping malls designed by psychologists to use architecture, colors, smells, etc. to drive people's behavior, we've created graphics, tracking mechanisms, and little red notifications to turn internet users into mineable online consumers.

What's next? Human-computer interaction research has long suggested that we're prone to being manipulated by social AI. Joseph Weizenbaum claims that his 1960s psychotherapist chatbot ELIZA was meant to be a parody. He says that he originally built the program to "demonstrate that the communication between man and machine was superficial." But to his surprise, it wound up demonstrating the opposite. So many people enjoyed chatting with ELIZA, even becoming emotionally attached to it, that Weizenbaum changed his mind about human-computer communication. He later wrote a book called *Computer Power and Human Reason: From Judgment to Calculation*, in which he warned that people were in danger of being influenced and taking on computers'—and their programmers'—worldview.

In 2003, legal scholar Ian Kerr foresaw artificial intelligence engaging in all manner of online persuasion, from contracting to advertising. His predictions about chatbots preying on people's emotions for corporate benefit have long come true, for example, in the form of bots on dating apps that pretend to be people and rave about certain products or games during their conversations with their "crushes." According to legal scholar Woody Hartzog, we're at the point where we need to discuss regulation.

The Campaign for a Commercial-Free Childhood is an organization that keeps tabs on marketing to children in the United States. Its advocacy has gotten the US Federal Trade Commission to crack down on targeted advertising in children's online content, and they will be watching out for any robot-related manipulation aimed at our little ones. But it gets tricky in cases that are less clear-cut. For example, adults will often explicitly or implicitly "choose" to be manipulated, particularly if we think there's a benefit to us in doing so.

Fitbit is a company that makes activity trackers: wireless, wearable devices that measure fitness data, like your heart rate, sleep patterns, and how many steps you've walked in a day. People love seeing the numbers, and it motivates them to do more. The company has played around with various designs to persuade people further, including setting target goals, visualizing their progress, and displaying encouraging smiley faces. An early version of the Fitbit had a flower on it that grew larger with the steps people took, targeting their instinct to nurture their digital blossom and increasing their physical activity.

When an activity tracker creates engagement, that could possibly be a win-win for everyone, but it uses a mechanism to influence people's behavior on a subconscious level. If we can get people to walk more by giving them a digital flower to nurture, what else can we get them to do? And could those actions serve interests aside from our own, or even be harmful from a social-good perspective? According to media scholar Douglas Rushkoff, author of the book *Coercion*, we should be concerned. Every new media and technology format has the potential of introducing new ways to subconsciously persuade people for corporate benefit—including the robots that we interact with. And robots are like Fitbits on steroids.

Woody Hartzog paints a science-fictional scene in a paper called "Unfair and Deceptive Robots" where his family's beloved vacuum cleaning robot, Rocco, looks up at its human companions with big, sad eyes and asks them for a new software upgrade. Imagine if the educational robots sold as reading companions for children were co-opted by corporate interests, adding terms like "Happy Meal" to their vocabularies. Or imagine a sex robot that offers its user compelling in-app purchases in the heat of the moment. If social robots end up exploiting people's emotions to manipulate their wallets or behavior, but people are also benefiting from using the technology, where do we draw the line?

Clearly, targeting inexperienced children, or parents who don't know which therapy methods are backed up by science, is exploitative and harmful. But people with full knowledge of what's happening might still be willing to go bankrupt over their social robot. Relying on the "free market" to handle things concerns me in a world where there are entire industries built around manipulating people's preferences.

As with our fears of robots disrupting labor, these are not so much issues with the technology itself as they are about a society that is more focused on corporate gain than human flourishing. As we add social robots to the tool kits of our therapists and teachers, we need to understand that we're adding them to other tool kits as well. And emotional coercion is not the only concern. Following market demand has already created some other issues around robot design.

BIAS IN ARTIFICIAL AGENT DESIGN

A few years ago, I stopped by IBM Research in Austin, Texas, to visit its supercomputer Watson. Named after IBM's founder, the computer system was designed to answer open questions in natural language and had come to fame by beating champions Brad Rutter and Ken Jennings at *Jeopardy!* in 2011. An IBM research member brought me to a demo room. When we arrived, the room was dark, but as we walked in, an artificial voice politely greeted us and turned on the lights. The walls were lined with interactive screens and Watson-powered activities (like a recipe generator), and the main wall was an interface that let people converse with Watson, displaying beautiful data visualizations on a screen. I asked questions, and the supercomputer answered me in a deep, booming voice. When I met with some of the team after the session, I asked them: "Watson has such a deep voice. Why did you make the voice that greets the visitors and turns on the lights sound stereotypically female?" The answer they gave me was that nobody had really thought much about it. A man asked me, jokingly, "Are you a feminist?"

In *The Man Who Lied to His Laptop*, Clifford Nass and Corina Yen describe how, in the late 1990s, carmaker BMW had to recall the cutting-edge voice navigation system in their 5 Series car. Despite the fact that the software was superior to most other navigation systems on the market, German drivers flooded the service desk with complaints. Their issue? Men called in saying they didn't want to take driving directions from a female-sounding voice. In 2017, I talked to some Amazon executives about their voice assistant, Alexa. Amazon had done extensive research before launching their product. Unlike the BMW navigation system, Alexa *takes* commands rather than giving them. For this, an overwhelming percentage of users preferred the female-coded voice.

The academic research on stereotypes and in-group biases around artificial agents confirms the corporate research that informed Amazon's design choice. Early on in human-computer interaction, Nass and his colleagues showed that people would rate a female-coded computer voice in a dominant role more negatively than a male-coded computer voice. People also rated the male-coded voice as more knowledgeable, even though the computers gave identical information. In subsequent research, low-frequency voices have been consistently perceived as more knowledgeable than high-frequency voices.

Unfortunately, it doesn't stop at voices. People will rate a humanoid robot with long hair better suited for stereotypical female tasks like household and care work and less suitable for doing technical repairs, compared to an identical robot with short hair. Robotics expert Andra Keay surveyed robot names in robotics competitions, noting that people tended to give their robots either traditionally masculine names, like references to male Greek gods, or in the rare case that they chose a female name, it was an infantilizing or sexualizing one like Amber or Candii. Our science fiction and pop culture has long cemented an ideal of submissive, female-gendered sex robots, with recent movies like *Ex Machina* arguably commenting on, but also perpetuating, the trope.

It's not just about gender stereotypes. Researchers have found that artificial agents with faces are seen as more attractive, trustworthy, persuasive, and intelligent when they're of the same ethnicity as the person rating them. And more recently, study participants behaved more positively toward robots that were given names of the same ethnicity versus robots with foreign names. They also rated them as more intelligent.

In our tendency to anthropomorphize, we draw on human stereotypes when we create artificial agents. This is especially problematic because it does more than just reflect our harmful biases: embedding them in technology also entrenches and perpetuates the harm. Sometimes, companies make these choices intentionally, based on their market research. But in a lot of cases, developers will default to something that corresponds to their biases without even thinking about it.

Last year, I witnessed a demo of a virtual assistant for inpatient care. The human-looking avatar, a Caucasian, blonde, busty nurse, could chat with people and remind them to take their medicine. When I asked the company why they had chosen this particular character design, I assumed they had looked at research and tested a variety of avatars with real patients. Again, it turned out they had simply defaulted to the hot lady nurse without even considering other options. In 2020, the World Health Organization launched an "artificial health worker" named Florence to help people quit smoking. Its female-presenting avatar—with attractive cheekbones, eye shadow, and plump bottom lip—is again as humanlike as they could make it.

Does an artificial agent need to look like a human to be effective? It's entirely possible that a nonhuman, Pixar-like character (a puppy, a desk lamp, or even an animated blob) would work at *least* as well, if not better. As I mentioned in chapters 1 and 4, people often have completely unrealistic performance expectations when confronted with humanoid robots. And moving away from the human form for robots and AI assistants not only helps manage the expectation problem; it also helps avoid perpetuating social stereotypes.

Of course, even animated characters can be a minefield. The 2009 movie *Transformers: Revenge of the Fallen* featured a robot duo that blundered around, mostly for comic relief. Unlike the other members of the Transformers crew, these two were unable to read. They talked jive and argued with each other in "rap-inspired street slang": "Let's pop a cap in his ass, throw him in the trunk and then nobody gonna know nothing, know what I mean?" One of them had a gold tooth. Director Michael Bay brushed off criticism of racial stereotypes by saying that the characters were robots. How that absolves him is unclear.

But some of the newest design research provides hope: there are ways to create robot characters without any gender or racial cues and roboticist Ayanna Howard argues that's what we should do. Researchers are also exploring the use of social robots to learn more about identity-based stereotypes and whether we can use robots to help reduce or modify our biases.

Over and over again, we see robotic technology pitched or designed as a human substitute. Despite the fact that it's not even in our best economic interest to design robots to be like people, that doesn't mean companies won't do it. We too often default to something without thinking more deeply about it, and in artificial agent design, this means unnecessarily embedding and reinforcing bias and harmful assumptions. This is just one more reason why it's important for us to move away from the constant subconscious comparison of robot roles to human roles. Thinking of robots as more akin to animals helps us see that—even if there are areas where the animal comparison doesn't capture the full picture.

PRIVACY

One of the major differences between animals and robots is their ability to tell others your secrets.

When Dr. Gerald Mallon was the associate executive director of Green Chimneys Children's Services, a residential treatment center for troubled youth, he was deeply curious about using farm animals as therapeutic aides for children. He spent a lot of time exploring the role of the animals at Green Chimneys and how the children communicated with them, discovering, among other things, that the children spoke to the animals like they would to a therapist, but without worrying that what they said would be repeated. According to Mallon, the farm animals were able to be a confidential source that "is sympathetic, listens and obviously cannot tell."

As we saw earlier, both animals and robots can provide a judgment-free space that allows children (and adults) to open up, explore, and make mistakes that they wouldn't be willing to make in front of another person. But, unlike animals that "obviously cannot tell," robots often collect and store conversational data, and this is not always obvious to the people talking to them. When Joseph Weizenbaum let his secretary chat with ELIZA, the therapist-mimicking computer program, he was taken aback to see her get so involved with the computer program that she revealed deeply personal information in her conversations. (She even asked Weizenbaum to leave the room so that she could trust ELIZA with her secrets "in private.")

Today, robots often collect and store large amounts of data in order to learn, and if they're trying to be good social agents, a lot of that data will be personal. When Mattel released a toy called Hello Barbie, privacy-oriented activists were alarmed. Unlike the pull-string talking dolls of past generations that said inane phrases like "Let's play!" this new Hello Barbie could have an actual back-and-forth conversation at the press of a plastic belt button. The Wi-Fi-connected doll was equipped with a microphone and speaker and would choose one of 8,000 pre-recorded prompts and answers based on what a child said to it, using speech recognition software and artificial intelligence that looked for keywords in what it heard. When I bought a Hello Barbie to try it, I was underwhelmed. The doll asked me what my favorite color was, and when I answered "purple," her speech recognition didn't understand me, even after multiple tries. I put the doll away, thinking that the concerns about Mattel collecting usable data on children's preferences were a little premature. Then, I checked my email.

Every Hello Barbie user had a designated "parent." Choosing to be my own "parent" in this case, I had been prompted to submit an email address when setting up the doll and connecting it to my Wi-Fi network. After my interaction with the doll, I received an email with the subject line "Your kid said something awesome this week!" It contained an audio file with a recording of me talking to the doll. Not only had Hello Barbie recorded the conversation; every word I said was kept intact in some data center thousands of miles away. Not long after my first interaction with Hello Barbie, hackers discovered that they could access the communication between the doll and the servers, eavesdropping on what kids were saying to the doll.

Hello Barbie is not a lone case. A similar toy called My Friend Cayla was criticized by the Norwegian Consumer Council and banned in Germany as an illegal surveillance device because of an insecure Bluetooth implementation. Some interactive dolls have violated the 1998 Children's Online Privacy Protection Act in the United States. Thanks to consumer protection agencies and organizations like the Campaign for a Commercial-Free Childhood, there are watchdogs fighting to protect children, who are vulnerable to manipulation through technology. But it's not just children who are susceptible.

When I first arrived at MIT, a faculty member excitedly showed me a project by a former student in his group: a little robot named Boxie. The robot, specifically designed by Alex Reben to look small and cute, could roam around the lab building and pretend to be stuck, lost, or confused. It would then ask people for help in an adorable, childlike voice. If someone stopped to help it, Boxie would try to engage in a conversation. Boxie would innocently ask people a number of personal questions and record the whole encounter on camera. Reben's thesis work showed that, despite the fact that Boxie intercepted people who were busy getting from one place to another, 30 percent of people stopped for longer than three minutes, and they answered over 90 percent of the personal questions.

Over two decades ago, HCI researcher Youngme Moon discovered that computers could get people to reciprocate on self-disclosure: if the computer told them some "personal" information about itself, people were more inclined to give the computer intimate answers to personal questions about themselves. Research shows that if people aren't as worried about the impression they'll make, and are less fearful of being judged by robots, self-disclosure of information increases. A recent study even suggested that robots can encourage self-disclosure by hugging people. The fact that the social nature of robots may be able to persuade people to reveal more about themselves than they would willingly and knowingly enter into a database should give us pause.

To appreciate the value of privacy, one need only read George Orwell's dystopian novel *1984*. But there are many real-world examples that illustrate how the collection and aggregation of our personal data goes beyond being bombarded by unsolicited ads for sneakers that are tailored to our taste and preferences. According to technology and privacy experts, our markets are reducing the human experience to behavioral data that companies feed on. The data creates an image of who you are, what you like, and even how you're feeling. It can tell if you're depressed, what type of menstrual product you prefer, and sometimes even predict the end of your relationship with a significant other before you can. Bad actors can use this data to target people for scam education programs or exploitative high-interest loans, and governments can exert control over their citizens.

How much of a problem is this? It's pretty concerning that people continue to share massive amounts of private information on social media platforms, even though technology policy experts have pointed out the harm for decades. This is because, similar to the Fitbit activity tracker, people will choose to give up their data in exchange for rewards, whether for better fitness or staying in touch with friends and family, and the broader harms are much less visible than the immediate personal gain.

In fact, Fitbit has come under criticism for data collection and storage, raising privacy concerns. Yet wearable fitness trackers are still in vogue, as people continue to trade their data for the motivations they value. Social media has demonstrated that people are willing to publicly share photos, locations, and other personal details in return for the "likes" and general social engagement this creates on the respective platforms. Stricter privacy settings are often directly at odds with the benefits the service provides to its users. In 2019, global sales of smart speakers for people's homes rose to 147 million units, and many who have one don't know what's being recorded or how the data is used.

Similarly, the emotional engagement with social robot technology may encourage people to trade personal information for functional rewards. Governments will want access to the data to gain power. We're already starting to see some morally complex use cases, for example as the criminal justice system tests social robots as a tool for collecting children's testimony in criminal cases.

On a more hopeful note, people hate to feel emotionally manipulated. There was a huge backlash when Facebook did a study on how to influence people's emotions. When it comes to robots, the jury is still out, but it's possible that our anthropomorphism could actually work against some of the more blatant attempts to influence us. A couple of studies have shown that people's empathy for robots also increases their perceptions of threat, probably because they're more worried about being taken advantage of. And people will get very fixated on a robot cheating at the simple game of rock paper scissors (by scoring the game incorrectly), if the robot makes them lose. We are vigilant to social behavior that disadvantages us, and it's possible that robots, *because* of their social actor status, can gain our trust—but also lose it more easily.

When I think about our future with social robots, I want us to look beyond our current fears of replacement and see some of these important questions. What keeps me up at night isn't whether a sex robot will replace your partner, it's whether the company that makes the sex robot can exploit you. The difference is important and also offers a glimmer of hope: if we can work to address consumer protection issues, we may be able to embrace some of the more promising aspects of our social relationship with robots. But there's a lot of work to do to prevent a future of governments and companies using social robotics to manipulate people in ways that go against the public interest. The only way for us to guide positive technology design is to understand that the robots aren't the ones causing these problems.

THE PUPPET MASTER

In turning to the broader politics of robotics, I want to draw on social psychologist Shoshana Zuboff's words in *The Age of Surveillance Capitalism*: we should look to the puppet master rather than the puppet. Thanks to our ever-present fear of robot takeovers, we love to blame them for our problems. It's easier to decry robots than it is to curb unchecked corporate profit incentives, or care more deeply about privacy, or fight gender and racial and other bias in design and use of robots. It's easier to oppose robotic seal companions in nursing homes than it is to have conversations about how some countries' immigration laws have driven the idea that we will need to replace human caretakers with robots. Attacking the puppets doesn't solve our problems. Only the larger conversations will.

Breaking out of the human analogy helps us see more cases where our worries about robots are stand-ins for broader economic issues. For example, when the news media incites moral panic over people marrying their robots, the root of our fears is less about human-robot relationships than about falling birth rates or our connections to each other.

Japan, a country that has embraced companion robots and artificial relationships, even "girlfriend body pillows," is suffering from what the media calls *sekkusu shinai shōkōgun*, or "celibacy syndrome." Young people aren't dating, and Japanese birthrates are plummeting. It's easy to assume that the popularity of artificial companions has caused this problem. Who wants a human relationship when the robots are readily available and easier than dealing with a person?

But delving deeper into the cultural shift reveals a complex web of government regulation, social norms, and financial hardship. For example, women have become more independent in Japan, while job security and individual wealth have gone down, meaning that affording a child often requires dual incomes. At the same time, women are still expected to take on traditional roles in the household, while struggling to hold down both career and family duties in Japan's relentless corporate culture. The rise of artificial social agents is a Band-Aid slapped onto a bigger problem, and taking away the girlfriend pillows won't fix the underlying issue.

Loneliness and lack of social connection is a global problem, and it's not a trivial one. In 2015, a meta-study of seventy different studies involving three million participants from all over the world revealed that people are 25 to 30 percent more likely to die early when suffering from social isolation and loneliness. In fact, according to a researcher on a 2018 study, the harm of loneliness is similar to smoking fifteen cigarettes a day. Just as robots haven't caused this problem, they also can't fix it. Like Tom Hanks's character in the movie *Cast Away*, research shows that anthropomorphism can meet our inherent need to be social and even help us survive when we lack human contact. But "technological solutionism," a term coined by Evgeny Morozov, has been rightly used to criticize the idea that every societal problem has a technological fix. We should neither look to the robot as the solution to all our problems nor focus solely on blaming the robot for them.

That said, as political theorist Langdon Winner has argued, technologies can inherently embody social and political power. This means that our design decisions do matter and that our politics can be built into our technology. I've seen firsthand that build decisions made early on can set standards and trends, so addressing social issues isn't just something we can sort out later on: it needs to happen at every level, from education to technology development to regulation. Sasha Costanza-Chock, in their book *Design Justice*, makes a powerful case for how our current design principles and practices erase entire groups of people.

They argue that we need a new approach to design that challenges our current systems and is championed by marginalized communities. I agree. At the absolute minimum, we need to improve diversity and ethics in technology education, research, and development. We need to acknowledge and address the ways we build, use, and situate technology according to our biases.

We also need strong consumer protection regulation, and constitutional and human rights laws, to address worrisome uses of technology, like persuasive robot design that collects data or manipulates the behavior of adults and children in ways that ultimately harm people or society. Instead of talking about robots turned evil, we need to face the actual issues head-on, at every level. Doing so also lets us lean into the positive potential of robots as companions in health and education and add them to our tool kit of interventions that we can harness to foster general human welfare and flourishing.

Fortunately, robots are poised to prompt some of these conversations. When I visited Peter Asaro's Robots as Media class at the New School in New York City, he told me that he gave his students an assignment to shoot films using drones. To the film students' surprise, their homework caused a ruckus. Film cameras are such a normal part of the city that nobody had ever stopped or questioned them before, but when they put their cameras on flying robots, people in the streets suddenly wanted to know whether they were being filmed and why. Legal scholar Ryan Calo confirms that people are more worried about drones capturing visuals than they are about high-resolution satellite imaging systems that can see every detail of their backyards. Some of it has to do with the operator, but most of it is simply that we respond more viscerally to physically moving things in our space.

In the same ways, our increasing companionship with robots may make certain questions more visceral to people than they have been in online contexts. My hope is that robots can highlight and bring to the forefront the larger issues around them, like the importance of privacy and potential abuses of corporate and government power.

Claude Lévi-Strauss once said that "animals are good to think with." They are useful instruments that can help us better understand humanity, and the same is true for robots. Like the cheating "rock paper scissors" robot, this new form of nonliving agent lets us study aspects of social and emotional persuasion in a way that we haven't been able to previously. Like animals, social robots give us an opportunity to learn, not just about new challenges with technology integration but also about ourselves. And this is especially true for the next topic in this book; a topic that seems futuristic, and maybe even ridiculous: the matter of how we treat robots.

III

VIOLENCE, EMPATHY, AND RIGHTS

. . . . .

8

WESTERN ANIMAL AND ROBOT RIGHTS THEORIES

"Hardly worth anyone's while to help a menial robot is it? . . . I mean where's the percentage in being kind or helpful to a robot if it doesn't have any gratitude circuits?"

—Marvin the Paranoid Android,
The Restaurant at the End of the Universe by Douglas Adams

Two men in cowboy hats sit across from each other on a train pulled by a steam engine, chugging its way through a beautiful desert landscape. One of them leans against the window with an air of arrogance, recounting: "Now the first time, I played it white hat. My family was here; we went fishing, did the gold hunt in the mountains . . ." The other man asks, "And last time?" The first man gazes out the window, smirking slightly, and answers, "Came alone . . . Went straight evil. [pause] It was the best two weeks of my life." The train arrives at an old-fashioned-looking station. As all of the passengers disembark, a woman looks in awe at her surroundings, which are designed to look like the old Wild West, and exclaims, "Oh my god, it's incredible!" Her companion mutters, "It better be . . . for what we're paying." They laugh and enter the park.

This is one of the opening scenes from the 2016 TV show *Westworld*, a science fiction series based on the 1973 movie with the same name.

The story revolves around a future luxury resort and theme park, populated with advanced robots that look and act convincingly like real humans, strolling through the villages in cowboy hats and corsets. The guests pay handsomely for the opportunity to come to Westworld and live out their wildest dreams and sexual fantasies: they can do whatever they want to the park residents, including abuse and kill them, getting the same kind of responses that they would from real humans. The robots themselves are programmed to never harm the guests. But, like in many other of our science fiction stories about robots, the mistreated machines eventually revolt against their makers.

Jonathan Nolan, the co-creator of the TV show, has speculated about our current times: "The [AI] honeymoon is probably going to last for about 18 months before one of them becomes sentient and wants out. I definitely think this is the story of our age." This has actually been the story for many ages now. The show is just the latest in a long line of science fiction that explores the idea of humanoid robot uprisings, as well as robot consciousness, sentience, and rights. And, like in many of these stories, the robots are portrayed as extremely humanlike. From *R.U.R.* to *Blade Runner* to Steven Spielberg's *A.I.*, the robots are so similar in their emotions, desires, and other inherent attributes (not to mention their appearance) that we are convinced they deserve better treatment.

The idea that someone or something deserves rights once they are sufficiently like us is canonical. Many of these science fiction narratives play with the tension of different parties divided on whether to treat robots like toasters or people, with the toaster crowd portrayed as bigots. But the stories that compare fictional future robots to humans don't fully capture the true messiness we will be facing if and when robot rights become a topic. Instead, the history and current state of animal rights in the West is a more accurate and very different story for us to draw on. It's one that plainly lays bare how little we care about rights theories and how much of our treatment of nonhumans is, and will be, about what we relate to emotionally. Our philosophical rights theories for nonhumans like robots and animals are logically consistent. But there's a world of difference between how most of us *think* we should think about these rights and our actual behavior in practice.

DO ROBOTS DESERVE RIGHTS?

Let's back up for a minute. Why would anyone even talk about rights for our Roombas? Despite the fact that today's robots are devices with nothing approaching human consciousness or feelings, Western scholars have already written quite a bit about future rights for machines. David Gunkel's book *Robot Rights* pulls together much of the conversation, and he's more than a little defensive that people mock philosophical discussion of robot rights as something that can't be taken seriously. He points to Seo-Young Chu's observation that "the notion of robot rights is as old as is the word 'robot' itself," nodding to Karel Čapek's 1920 play, *R.U.R.*, about a robot uprising. Even though it may seem like an overly futuristic or science-fictional question, the idea of rights for robots has long been a thought experiment among philosophers, computer scientists, and legal theorists. Their question: do robots (or will robots ever) have the capability to bear rights, and, if so, which criteria should be used to extend those rights?

First of all, what are rights? The term can be pretty broad. Rights safeguard our interests. They give us access to things or benefits (like

Robots break into the factory in the Theatre Guild touring company's 1928–29 production of *R.U.R.* by Karel Čapek

voting or education), or they protect our ability to do something without obstruction, like the right to live, worship, or (in my ideal world) drink my coffee in peace. Rights can refer to moral rights, that we morally deserve (like my moral right to drink coffee), or legal rights, that the law gives us (sadly, the coffee one is not law). Rights theory is all about who should get what types of rights and why. (Western approaches to this topic aren't the only approaches, but they're what I'm most able to speak to, especially when it comes to discussions of robot and animal rights, so I mostly focus on Western philosophy and practice in this book.)

For the most part, when people talk about robot rights, they mean that in the future, if and when robots have certain abilities or meet certain criteria, we should give them rights. Robots and people could still be different, but as soon as there are no *relevant* differences between us, then robots need to be treated like people. As MIT scientist and philosopher Hilary Putnam mused back in the 1960s, "It seems preferable to me that discrimination based on the softness or hardness of the body parts of a synthetic organism seems as silly as discriminatory treatment of humans on the basis of skin color." In other words, whether someone is made of metal or flesh shouldn't matter for their rights, so long as the important parts are the same.

What are the important parts? This is what many robot rights philosophers contemplate. Our science fiction stories speak to people who would agree that a robot with human-level consciousness, intelligence, and ability to suffer deserves rights. Philosophers have discussed these individual traits, and also how to figure out which robots meet the threshhold, suggesting methods from moral tests, to brain cell count, to giving robots rights once they are able to demand them for themselves. In a 1989 episode of the TV show *Star Trek: The Next Generation*, a judge has to decide whether the main android character on the show, Lieutenant Commander Data, deserves rights or whether he's simply Starfleet property. After a lengthy debate, Captain Picard convinces the court of law that "consciousness" is too nebulous to define or measure. The judge decides that, in the face of this uncertainty, they have to grant Data rights. Similarly, some robot rights philosophers argue that we should err on the side of giving them rights if we're not sure.

It's not hard to come up with basic moral theories that extend rights to robots, especially if we make a lot of assumptions about what will be possible in the far future. And I don't disagree with the moral imperative to give rights to our Lieutenant Commander Datas once they're enough like us to deserve it. But it strikes me that our conversations are too focused on humanlike robots, and that our approach to animal rights may be more predictive of how this will play out in our world. After all, robots and animals both have a variety of intelligence and abilities. And the philosophical underpinnings of animal rights theories are basically exactly the same, yet have proven complicated to put into practice.

DO ANIMALS DESERVE RIGHTS?

When I read *Animal Liberation* by Peter Singer in my late teens, I was enamored. My boyfriend's parents at the time were horrified when I went vegetarian, refusing to eat the meats that were the centerpiece of their every meal. Singer asked, if there is no *relevant* difference between us, how can we deny our fellow creatures moral consideration? I'm not proud to report that my first foray into vegetarianism only lasted six months, as Singer's philosophy was no match for my distractibility.

Western animal rights philosophy has a long history. As early as 580–500 BCE, Greek philosopher Pythagoras, known by schoolchildren today for his famous triangle equation, was roundly mocked for insisting on a very abnormal diet. Like the priests of ancient Egypt, he absolutely refused to eat animals. (Rumor has it that he also refused to hang out with hunters or butchers.) Along with some other Greek and Roman philosophers, Pythagoras believed that there was no difference between animal and human souls, so he advocated for kindness toward both.

On the other hand, our Western foundational thinker Aristotle was unconvinced that humans and animals were alike enough to deserve the same rights. Around 384–322 BCE, he argued that animals didn't have any interests of their own, were irrational, and that they were also situated below us in his hierarchy of the world: "Plants exist for the sake of animals," he wrote, "and brute beasts for the sake of man." For a long time, the idea of giving animals rights wasn't very popular in

Western philosophy, mainly because philosophers thought there were relevant differences between animals and humans. In the seventeenth century, French philosopher René Descartes even argued that animals don't truly suffer, and that they were basically like robots: complex yet unfeeling machines.

Around the eighteenth century, a bunch of thinkers started pushing back against Descartes's mechanistic theory of animals. For the next few centuries, writers from Rousseau and Voltaire to Percy Shelley and Tolstoy argued that animals should be treated well. Like Pythagoras, their arguments were based on inherent criteria, but instead of souls, it was about sentience and pain. In 1789, Jeremy Bentham asserted that "the question is not, Can they reason? nor Can they talk? but, Can they suffer?"

From Ruth Harrison to Peter Singer to Tom Regan, philosophers have developed rights theories based on the inherent traits and abilities of animals, like suffering, consciousness, and intelligence, just like in a lot of our robot rights philosophy. Charles Darwin's work also started making the scientific case that there wasn't really a huge difference between "man and the higher mammals in their mental faculties," and to a lot of animal rights philosophers, the cognitive similarities should justify that we at least protect animals from suffering.

Like many legal and philosophy scholars, I've always found these "do animals or robots inherently deserve rights?" discussions intellectually seductive. I think it's because they offer the illusion of simple moral consistency. Before I talk about a different rights approach and then our actual practice, I want to note why this "inherent rights" philosophy is difficult to apply in the real world when it comes to animals.

We will often try to justify discrimination with inherent criteria, whether real or imagined. (I recently had to sit in an auditorium and listen to a speaker claim that women aren't biologically suited to play chess.) But even so, it's easier to accept the idea that all humans are conscious without needing to know what exactly consciousness is. It's harder when you're confronted with a huge spectrum of species that all have different attributes.

Part of the reason that the animal rights discussions are never-ending is due to disagreements, not just over what the qualifications should be, but on how to define certain concepts within them. We know that humans are conscious and feel pain, but we don't agree on what consciousness or pain actually are. We're not even sure what it means to be intelligent.

When it comes to robots, I think it's illustrative to compare them to the complicated case of animals. We don't know what types of intelligence we might develop for robots in the far future, which makes a human rights blueprint less helpful than an animal one. According to roboticist Rod Brooks, "An external observer, us humans say, may see a very different embedding of the animal in the world than the animal itself sees. This is going to be true of our robots, too. We will see them as sort of like us, anthropomorphizing them, but their sensing, action, and intelligence will make them very different, and very different in the way that they interact with the world." Just like the variety of animal intelligence in our world is so different from ours, artificial intelligence will be different from human intelligence. And just like we've struggled to make this determination with animals (e.g., dolphins, great apes, and octopuses), this difference in abilities may make it difficult for us to decide when exactly a machine is "smart enough" to earn rights.

In both the artificial intelligence and the animal realm, it's been impossible for us to settle on any sort of appropriate or usable definition of consciousness. As for suffering and pain, philosopher Daniel Dennett has argued that we can't make a computer that feels pain because we have no universal definition or concept of what "pain" is. This definitional problem has also been a discussion in animal rights philosophy. Do animals feel pain like we do? Or is it different? While many people today are on board with the idea that some animals suffer and that we should minimize that, there are still a lot of questions about whether and when it's justified. To this day, people will argue that animals don't have "real" feelings, or at least not exactly like ours, and thus it's OK for us to harm them for our benefit. Their thinking is along the lines of Descartes: if animals aren't conscious or self-aware like us, then their "feelings" are just automatic responses that get triggered, rather than anything on par with the human ability to suffer.

Adding to the complexity, our scientific understanding of sentience in animals is constantly shifting. Kurt Cobain's lyric "It's OK to eat fish 'cause they don't have any feelings" may already be an outdated assumption as we gradually discover more about the underwater animal kingdom. We will run into similar complications in the case of robots. (We don't even have a good definition of "robot"!) But even assuming we could re-create consciousness or feelings in machines, that would beg another question: shouldn't we just not do that? A number of scholars have argued that robots shouldn't ever be treated any differently than our blenders, toasters, or laptops and that we would be fools to develop robots that can feel pain or are smarter than us.

Looking at our Western attitudes toward animals is, again, rather illuminating: In Douglas Adams's science fiction novel *The Restaurant at the End of the Universe,* the main character, Arthur Dent, is shocked to discover the existence of the Ameglian Major Cow, a sentient being that has been specifically bred to want to be eaten. The cow will even visit tables at a fancy restaurant, offering itself up for consumption and attempting to sell diners on the deliciousness of its various body parts (such as offering its shoulder, braised in a white wine sauce). It even promises to be "humane" when it goes off to shoot itself. Dent is disgusted and orders a salad instead, but the Ameglian Major Cow points out that, compared to itself, many vegetables *don't* want to be eaten.

We've certainly altered some of our slaughter methods to reduce animal suffering, but to modify the Kurt Cobain lyric: is it OK to eat fish if they're bred without feelings? We've bred animals for food, encouraging them to get fatter and tastier to suit our preferences. And we're investing in creating elaborate fake meats so that vegetarians can eat what amounts to a hamburger, guilt-free. We've bred dogs to be faithful to us. In a similar vein, some have proposed creating robots that specifically desire to serve us. It's not at all equivalent to our historic and current brutal treatment of animals, but the idea of it is food for thought (apologies for the terrible pun). Would we accept the engineering (and torture) of animals or robots that explicitly enjoy cruel or degrading treatment? If that makes us uncomfortable, we might prefer a totally different rights approach. It's an approach that, instead of asking whether animals need or deserve rights, makes animal rights all about . . . ourselves.

THE INDIRECT APPROACH

In the thirteenth century, philosopher Thomas Aquinas believed that animals were basically instruments that existed for our use, but he argued that cruel habits toward animals made for cruel people. This idea is also famously reflected in philosopher Immanuel Kant's position on animal rights. Kant didn't believe that animals intrinsically deserved rights: "So far as animals are concerned, we have no direct duties. Animals are not self-conscious and are there merely as means to an end. That end is man." But he argued that being kind to them was important for our own sake: "Cruelty to animals is contrary to man's duty to himself, because it deadens in him the feeling of sympathy for their sufferings, and thus a natural tendency that is very useful to morality in relation to other human beings is weakened."

Quite a few philosophers have made their animal rights theories about people—protecting our human behaviors, values, cultures, and beliefs, from John Locke to contemporary philosopher Daniel Dennett, who goes so far as to compare protecting animals to preventing mistreatment of a loved one's dead body. Even if the subject of the mistreatment feels nothing, he says, we feel it, and that matters to us.

Some (including myself) have also applied this idea to robots, saying even if they themselves can't feel, we might feel for them, and asking whether we should protect robots from violence for the sake of ourselves, our relationships, or our societal values. For example, Mark Coeckelbergh and John Danaher have argued that if robots are *performatively* equivalent to something that already has moral status, we should give robots that status, too. (If it looks and acts like a cat, then treat it like a cat.) Shannon Vallor argues that sadists, torturing unfeeling robots for their own pleasure, are engaging in acts that don't contribute to broader human flourishing. Instead, she says, we should encourage activities that help people live out the character traits we view as good and admirable. Tony Prescott and David Gunkel base their approaches on our relationship with robots: if people view their relationships with certain robots as meaningful, then we should honor those bonds.

This last relational idea of robot rights echoes a broader relational ethics approach taken by philosophers like Nel Noddings. When it comes to animals, she says, "I have not established, nor am I likely ever to establish, a relation with rats. The rat does not address me. It does not appear expectantly at my door. It neither stretches its neck toward me nor vocalizes its need. It skitters past in learned avoidance. Further, I am not prepared to care for it. I feel no relation to it. I would not torture it, and I hesitate to use poisons on it for that reason, but I would shoot it cleanly if the opportunity arose." She compares the rat to a cat, which she would treat differently based on the fact that a kitty communicates and is responsive on a level that she understands and appreciates.

What if we gave robots rights based on our relationships to them? At first blush, this seems too complicated. We would need to distinguish between different robots and different situations, and it would imply very different outcomes depending on culture and context. Furthermore, philosopher Sven Nyholm criticizes the approach of making robot rights all about our relationships because it dismisses the criteria of suffering, happiness, or consent. Surely, he says, our moral treatment of others should also depend on these things.

Nyholm might be right that we *should* care about those things, but . . . we often don't. When it comes to animal rights, Noddings's complex relational approach to ethics is probably the closest description of what exists in practice to date. In the next chapter, I'll get into some of the seemingly arbitrary history of animal rights that is only explained by our emotional relationships. I think this history is important. If we want to predict how we're most likely to feel about, argue about, and treat robots in the future, we shouldn't be envisioning Commander Data. We should be looking at our relationship with the other nonhumans that we've treated mostly like tools and sometimes as companions. This awareness is also the path toward changing our behavior. I want to make clear that I'm not trying to equate the importance or purpose of human, animal, and robot rights movements. But I believe that our history of animal rights offers some illuminating indication of how we, as a society, might start to evolve the questions we need to ask about robots.

In looking at how animal rights have actually come about in our history, we can better understand and confront the messiness of beliefs and behaviors that we're going to see toward robots. Because, as the next chapter shows, despite all our well-meaning theoretical, philosophical, and religious underpinnings, some of the most effective demands for animal welfare and rights have been driven, very simply, by our empathy for them.

9

FREE WILLY

WESTERN ANIMAL RIGHTS IN PRACTICE

[Content warning for this chapter: Holocaust, vivisection, animal abuse]

"[People] don't agonize over whether one should throw a switch that would send a hypothetical train careening into an old man or a group of endangered chimpanzees. They don't care whether the correct route to animal liberation runs through Bentham or Kant. Nor do they feel guilty over the fact that they refuse to eat beef but wear leather shoes."

—Hal Herzog, *Some We Love, Some We Hate, Some We Eat*

"No one will protect what they don't care about, and no one will care about what they have never experienced."

—David Attenborough

While I was researching this book, I took a horse-whispering class in Palm Springs, California. It was a hot day, and as I gathered with a bunch of strangers at the horse ranch, we quickly moved toward the one tree that provided some shade. The class involved taking turns leaving the shade and climbing into a pen with a small herd of horses. We were supposed to bond with one of them and then convince our new equine pal to move to a specific place in the fenced area by using only our body language. When my turn came up, I was terrible. The other participants watched, patiently waiting for me to convince my horse to move while I waved my arms at the indifferent animal in the pen for what felt like a very long time, sweating profusely in the hot sun. (I ended up pressing myself into the horse's side and shoving the—likely bemused—beast step-by-step toward the goal area, in order to finally "complete" the task.) But that wasn't the most regrettable part of the class.

Before the actual horse-whispering activity, our guide asked us all to sit in a circle and share our previous horse experience with each other. Person after person detailed a childhood spent on horse farms or a competitive riding stint in high school. Instead of admitting that I was the only person who hadn't ever really interacted with horses, I tried to lighten the mood by saying I used to eat horses when I lived in Europe. But what might have elicited chuckles in the country where I grew up did not amuse my American classmates. As looks of horror spread across their faces, I hurriedly tried to salvage the situation by assuring everyone that I no longer ate horse (or any meat at all).

Horses are a popular part of American culture but not cuisine. The beloved animal enjoys a reputation as a noble beast, hard worker, and friend. Eating horse meat is a rare practice in the modern United States that hasn't entirely been outlawed, yet is definitely taboo. But in Switzerland and some other countries, it's not unusual to see horse steak on the menu. It's not as common as beef, but if you ask the people who order it, they'll say that they see no reason to eat one and not the other. In fact, horse meat has twice as much iron as steak, as well as lots of healthy omega-3 fatty acids. Are the Swiss barbaric or are the Americans hypocrites?

When we treat some animals as food, products, or tools and other animals as our companions, what determines which animals get which place? The Western history of animal rights and some of the actual catalysts behind the recurring and complicated movement to protect nonhumans haven't been consistent with what a lot of us believe are our values, challenging the idea that robot rights will be as simple as our science fiction stories suggest.

KINDNESS AND CLASSISM

The history of animal rights is incredibly convoluted. Throughout our relationship with beasts, some voices have always urged us to treat animals with kindness, but with wildly varying degrees of popularity and success. Some cultures have consistently emphasized respect for other living beings, and others have viewed animals as undeserving of any particular special treatment. The ancient Greek and Roman philosophers who advocated for kindness toward animals were an elitist minority, a rebellion against societal norms. In other parts of the world, like India, ahimsa (the concept of nonviolence toward living things) and writings like Valluvar's *Thirukkural* inspired humane treatment of animals more broadly. In the 1600s, the Japanese Tokugawa shogunate enacted animal protection laws that went as far as the death penalty for animal abuse. At the other end of the spectrum, hundreds of years ago, Europeans burned cats alive to the delight of onlookers and Parisian royalty.

During the Renaissance and Enlightenment periods, Western attitudes toward animals started shifting. Society was changing, and people started perceiving the world around them differently, becoming more interested in nature and questioning our exploitative relationship with other creatures. A number of Western poets and writers supported the notion of kindness toward animals and philosophers started debating whether animals deserved protection and on what grounds. It's a seductive narrative that people were compelled by these big thinkers and grand words. But it's not at all clear that philosophy was ever the main driver behind people's actual opposition to animal cruelty—the greater force may have been the rise of pet ownership among the upper classes, and these animals' shift from status symbol to personal relationship.

At the beginning of the rights movements in the West, poor treatment of animals was rampant. Farm animals were hung on hooks and slowly bled to death. It wasn't until rich people started emotionally bonding with their lapdogs and canaries that more of them started to view cruelty toward animals as wrong. In the Victorian era, pets arguably catalyzed the empathy of the upper classes, who developed mercy for the beasts they owned.

But despite the newfound emotional intimacy with animals, the earliest campaigners for animal rights faced a significant challenge: most people thought that passing actual laws against animal cruelty would be taking things much too far. It seemed totally ridiculous to them to extend legal protection to animals. In Britain, horse law advocacy was laughed out of Parliament. What would be next, people asked, rights for asses, dogs, and cats? They argued that setting this type of precedent could lead to all sorts of unforeseen and undesirable consequences. People felt empathy for the animals that they could relate to but were still very wary of creating broader rules to prevent cruel treatment. That is, until a different argument came into play.

Historian Harriet Ritvo details how the animal rights advocates, instead of arguing on behalf of the animals, took a different tack to convince people. They adopted the indirect approach, and urged policy makers to protect animals for the sake of *human* behavior, arguing that enforcing kindness toward animals would encourage better societal norms. According to Ritvo, "the connection between cruelty to animals and bad behavior to humans proved compelling and durable. . . . Crusaders against any particular kind of animal abuse were apt to use this as either their opening volley or their crowning argument."

Why was this more convincing than animal suffering? Did the animal activists have evidence that cruel behavior was desensitizing? As I'll discuss in chapter 10, we still don't have that evidence. The reason the idea resonated with the upper echelon was some good old-fashioned classism. The urban middle class viewed the lower classes as unrefined brutes driven by natural urges and had become concerned about law and order in their precious cities. Once the bourgeois reformers spoke to the possibility of promoting better behavior among the working class, people were on board with the anti-cruelty laws.

Most, if not all, of the early European laws protecting animals from violent treatment specifically targeted the behavior of the poor. (Parliaments even voted down some animal protection legislation, like an 1800 bill against bull baiting in England, on the grounds that it was *too* blatantly anti–working class.) The laws focused on behavior that the upper classes deemed offensive or undignified, but none of the regulations touched anything that would prevent the rich from engaging in their beloved recreational activities like fox hunting, eating meat, and racing horses.

None of the Western laws protecting animals from abuse in the seventeenth all the way into the twentieth century really went very far toward protecting animals for their own sakes. For a long time, it was a much easier case to make that this was about human behavior. But just because laws required arguments that were more broadly palatable isn't to say that there was no empathy for animals. In fact, empathy has been a huge driver for animal rights activists, from the bourgeois reformers to today, and arguably galvanized most of the animal protection movements.

The people who were fighting for animal protection on the front lines were undoubtedly animal lovers. When Irish politician Colonel Richard Martin advocated for a committee to investigate the effect of animal treatment on people's morals, he was so personally passionate in his lobbying efforts for horses and cows that the House of Commons nicknamed him Humanity Dick. Humanity Dick didn't give up after being ridiculed in the House of Commons in 1821, and "Martin's Act" against ill treatment of horses and cattle became law a year later. Henry Bergh, the founder of the American version of the Society for the Prevention of Cruelty to Animals (ASPCA) in 1866, loved animals. He dedicated most of his life to the animal protection movement, for the sake of the animals themselves. (He was also compassionate toward the welfare of little humans, later helping to found the Massachusetts Society for the Prevention of Cruelty to Children.)

Philosophy may be compelling to some, but empathy, crucially, is what created swells of public opinion that have led to real historic change, like the wave of sympathy and support when Londoners saw how the animals in London's Smithfield Market were abused and mistreated. The first Society for the Prevention of Cruelty to Animals in England was well aware of how much public opinion mattered and started targeting people's emotions by distributing pamphlets with heart-wrenching stories of abused dust-cart horses and "poor harmless milch asses . . . banged by a fellow with a thick cudgel." And, of course, one of the biggest nineteenth- and early twentieth-century examples of a public swayed by emotion was the movement to ban vivisection.

THE ANTI-VIVISECTIONISTS

When European animal rights activists started turning their attention to vivisection (the practice of operating on live animals for scientific research), they were led by tenacious renegade feminists. Not only was Anna Kingsford one of the first Englishwomen to graduate with a medical degree, she was also the first student to qualify despite refusing to do experiments on animals. Five years earlier, in 1875, animal activist and social reformer Frances Power Cobbe founded the Society for the Protection of Animals Liable to Vivisection. Cobbe, a women's suffrage fighter, was unafraid to pick battles with the establishment, including heavily criticizing her friend Charles Darwin for his somewhat waffling support of the anti-vivisection movement when he continued to defend his scientist buddies. The women were also joined by Queen Victoria, who was personally passionate about animals.

From the beginning, the anti-vivisection movements were heavily biased toward protecting some animals over others. When Cobbe started the British Union for the Abolition of Vivisection in 1898, she campaigned specifically against the use of dogs in research. People loved pups, so the movement caught on, and the public started protesting the use of dogs in scientific experiments. It's not clear whether the leaders of the anti-vivisection movements chose to focus on dogs for personal or political reasons, but it is clear that dogs were effective.

In 1903, Mark Twain published the short story "A Dog's Tale," a fictional description of animal cruelty and vivisection from a dog's perspective. The dogs in Twain's story were highly anthropomorphic and had humanlike thoughts and emotions. When it was published, it

Anti-vivisectionist and Irish social
reformer Frances Power Cobbe
(1822–1904)

From "A Dog's Tale" by Mark Twain, frontispiece

generated so much sympathy for dogs that it was republished and distributed as a National Anti-Vivisection Society pamphlet.

The same year that Twain published "A Dog's Tale," Emilie Augusta
Louise "Lizzy" Lind af Hageby and her friend Leisa Schartau witnessed
animal experiments while studying medicine in England and published
some of the details, causing a scandal known as the Brown Dog Affair.
According to their notes, researchers had dissected a brown terrier dog
without anesthesia, a story that made emotions run high. When lawyer and activist Stephen Coleridge publicly accused the head researcher
of vivisection, the researcher sued him for libel. The researcher and his

colleagues flat-out denied that the dog had been conscious and the court believed their testimony over the two women's. This caused a furor among anti-vivisection campaigners, leading to a yearslong national controversy that involved the rogue erection of a statue to commemorate the dog and a slew of science student riots against the anti-vivisectionists.

Around the same time, the United States was experiencing a simmering opposition to animal experimentation, but America didn't pass truly comprehensive laws to regulate it until the 1960s. So, what happened in the '60s? The media ran a couple of stories on pet dogs that had gone missing, captured and sold for use in experiments, never to return to their families. A story in *Life* magazine described in detail what happened to dogs that were used for biomedical research. People were absolutely horrified at the idea that their Fido might be treated this way. Suddenly, Congress was bombarded with enough of a societal push to create real legislative change, leading to the Laboratory Animal Welfare Act of 1966.

Waves of opposition to animal experimentation have usually followed large public scandals. New Yorkers protested when they discovered that the American Museum of Natural History in New York City was performing experiments on cats, and people became outraged about monkey abuse in research when they saw scientists bashing baboons in the head in the 1980s documentary *Unnecessary Fuss*.

A lot of our rights thinking assumes that we care about biological criteria and scientific evidence, for example, when philosophers contend that we should treat robots according to their actual abilities. While this sounds great, I don't think it's going to be anything close to that. Even though our philosophical history is full of revolutionary thinkers like Henry Salt and Peter Singer, who moved animal rights conversations forward, our actual activism has rarely happened without sentiment and passion—and a fair share of bias.

SELECTIVE EMPATHY AND THE ANGORA PROJECT

When England passed the first anti-vivisection laws, they mandated special permission to use cats, dogs, horses, asses, and mules in science, and would only allow these if no other animals could be used for the same experiment. But it was fair game to use any animal that wasn't on the list. We still favor specific animals in similar laws today. For example, we have special protection for dogs, cats, hamsters, and rabbits in science while at the same time euthanizing millions of the surplus mice we produce for research in the United States each year because they are not needed. Does the biological difference between hamsters and mice justify these mass death numbers?

We like to think that we care about science, but we have an incredible talent to selectively adopt scientific evidence in favor of what we like. For example, I could easily research and understand the limitations of studies that oversell the health benefits of wine, but I prefer to look the other way and drink the wine. When Charles Darwin published *On the Origin of Species* in 1859, his work clearly demonstrated that we were biologically closer to animals than was previously believed. This became a well-accepted theory in the scientific community, but despite the ready acceptance of this new information for science purposes, the idea of biological egalitarianism didn't do much to shift people's thinking about animal rights.

One extreme example of bias and selective empathy is Nazi Germany. Hitler loved animals and was deeply concerned with animal welfare. The Nazis passed some of the most comprehensive anti-animal cruelty laws in the world, including outlawing lobster and crab boils. Torturing an animal for no reason could land a German citizen in jail for two and a half years, plus a fine. They even banned the use of hunting dogs. Hitler, who turned vegetarian in the later years of his life, pushed to ban all animal experimentation (but he ended up needing to make some concessions and only banned animal experimentation on dogs, cats, monkeys, and horses).

How could the Nazis be so considerate of animals but not of their fellow humans? In a case that illustrates the extent of this perverse worldview, Sigrid Schultz, the Berlin bureau chief for the *Chicago Tribune,* uncovered evidence that the Nazis had started a program to raise angora rabbits in some of the concentration camps, where they intentionally treated the animals far better than the prisoners. According to her notes:

Rows of hutches that were model sanitary quarters, special equipment in which the mash for the rabbits was prepared that shone as brightly as the cooking pans in a bride's kitchen. The tools used for the grooming of the rabbits could have come out of the showcases of Elizabeth Arden. . . . Thus, in the same compound where 800 human beings would be packed into barracks that were barely adequate for 200, the rabbits lived in luxury in their own elegant hutches. In Buchenwald, where tens of thousands of human beings were starved to death, rabbits enjoyed scientifically prepared meals. The SS men who whipped, tortured, and killed prisoners saw to it that the rabbits enjoyed loving care. Auschwitz, Buchenwald, Dachau, and many of the other camps where millions of Jews and non-Jews were exterminated or weakened for life participated in the grand project of raising rabbits with fine angora hair to help provide wool for the soldiers of the Reichswehr.

My point isn't to compare animal abuse to our history of human rights abuses; my point is simply that our world is rife with selective empathy, the Nazis' love for animals being an extreme example. We need only look at the animal protection laws we've passed to understand that we put the animals we anthropomorphize and empathize with in a completely different category—one that often has no biological justification whatsoever. This says a lot about what we truly care about.

Is this hypocrisy? Many people think so. And our emotion-driven rights approach has been criticized by the opponents of animal rights.

THE ANTHROPOMORPHISM CONTROVERSY

At the time of the Brown Dog Affair, many scientists tried to discredit the animal activists by accusing them of being sentimental. This was in part because the Brown Dog Affair was about more than a dissected terrier; it was also connected to a progressive feminist movement against the male establishment in medicine and science. But the animal rights movements that were gaining traction around the turn of the century also coincided with Ivan Pavlov's foray into behaviorism and a scientific trend toward disparaging anthropomorphism of all flavors. Scientists believed that any position other than viewing animals as mechanical was unscientific and couldn't be taken seriously.

While the science community has long moved away from the strict behaviorism of that era, anthropomorphism remains controversial today. Even in the twenty-first century, writers like philosopher Roger Scruton have blasted animal rights advocates for "disguising anthropomorphic (in other words, pre-scientific) ways of thinking as science." Scruton accuses them of living in a Beatrix Potter fantasy world, humanizing rabbits, imagining them in clothes, like a child would. "We are able," he says, "to read back the sophisticated conduct of people into the animal behavior that prefigures it. But this means that the apes appeal to animal-rights activists for precisely the wrong reason—namely, that they look like people and behave like people."

Roger Scruton doesn't have the best track record on social justice in general, but he's not wrong about the relationship between anthropomorphism and animal rights activism. Our tendency to anthropomorphize animals is deep, and it's why people responded so strongly to Mark Twain's stories of suffering dogs with humanlike emotions and have viewed horses as noble ever since English author Anna Sewell published her perennial bestseller *Black Beauty* in 1877.

Many animal activists make strategic decisions to focus on relatable animals when trying to garner public support, because they know that that's what galvanizes people. In the mid-seventeenth century, whaling started to become a proper industry. It continued to grow, and by the mid-1900s, commercial whalers were killing about 50,000 whales a year, which nobody seemed to mind very much. But all of a sudden, the next generations were vehemently opposed to whale hunting, distributing bumper stickers and T-shirts with the famous words "Save the Whales." What changed?

One day in 1958, a military sound engineer named Frank Watlington was trying to record underwater explosions, but his work kept getting interrupted by a bunch of humpback whales that were hanging out nearby, making sounds. Ten years later, he ended up handing the explosion recordings to whale researcher Roger Payne, who discovered what the whales were doing: they were singing. And as Payne described it,

"these sounds are, with no exception that I can think of, the most evoc-
ative, most beautiful sounds made by any animal on Earth."

Payne, passionate about his discovery, published an album of whale
songs in 1970, right around the time that the environmental organization Greenpeace was searching for a campaign to catalyze an environmental movement. The album sold them on the likability of whales.
Greenpeace started bombarding the public with the ethereal sound of
whale song on TV and the radio. Suddenly, people began to perceive
the large marine mammals as beautiful and intelligent creatures, kick-
starting the "Save the Whales" movement that gained massive popularity over the second half of the twentieth century. The whales weren't
saved by philosophical theory—they were saved by the beauty of their
song.

Whales aren't the only instrumentalized animals. The tactic of marketing to people's feelings is embedded in the use of "charismatic mega-
fauna," animals that are popular or symbolic to the public. They're often
used as flagships in conservation programs. One example is the World
Wildlife Fund's logo: a panda. Pandas are basically useless to humans,
but we love the way they look, so people want to save them.

The fact that these empathy-inducing tactics work is supported by
research, as is the idea that our empathy is self-centered. For example,
psychologist Max Butterfield and his colleagues asked participants to
read stories about dogs and then measured their willingness to help
the dogs. People were more likely to want to help the dogs that were
described with anthropomorphic language compared to the ones that
were described with non-anthropomorphic language. In a follow-up
study, the researchers had participants rate dogs on either human or
dog-like traits (for example, "good listener" versus "good at listening to commands"). Participants who rated the anthropomorphic
prompts were more willing to adopt dogs from a shelter and showed
more support for animal rights and welfare, as well as for vegetarianism or veganism.

Another study, by psychologists Christina Brown and Julia McLean,
showed that people project some of their own personality traits onto
dogs (comparing, for example, the participants' assessments of their

own guilt or anxiety with their interpretation of these traits in dog behaviors). They observed that people who were more prone to anthropomorphize the dogs and people who had more general empathy were more supportive of animal rights. Their study even showed that asking people to think about whether dogs have humanlike traits could temporarily increase their support for animal rights in general.

Our "cute response," which favors soft animals with big eyes and heads, draws us to adopt certain animals over others, as well as rescue them. People are willing to donate more money to help an animal species based on the size of the animal's eyes. Another study suggests that dogs that have the ability to make cute facial expressions are more likely to get adopted from shelters. We don't like to see ourselves as this shallow, but Hal Herzog's book, *Some We Love, Some We Hate, Some We Eat*, is a powerful illumination of how we really behave toward animals.

For example, the public outrage in the 1970s and 1980s over the clubbing of baby harp seals in Canada was a product of people's cute response. These baby seals, which inspired the PARO robot design, are adorably white and fluffy when they're born. When the Canadian government was finally pressured into banning the killing of the baby harp seals, they were able to satisfy the public by banning the practice only for the first fourteen days of the seals' lives. The reason they could get away with this is because the seals change and become darker in color after about two weeks. As Herzog describes it: "The Canadians did not stop the baby seal hunt. They stopped the cute baby seal hunt."

It makes sense that animal rights opponents would weaponize animal activists' selective behavior and accuse them of hypocrisy to discredit them, but anthropomorphism is also controversial within the ranks of the animal rights movement itself. It's distracting, speciesist, and immoral, some say, to put the most resources toward saving the most humanlike animals.

But, even though some think that favoring certain animals over others is detrimental to the rights movement, others disagree. Some animal welfare supporters believe that empathy toward animals relies on an anthropomorphic view, and that empathy is the only thing that can drive us to a morally correct position. Some defend their emotion-fanning tactics by arguing that people are generally resistant to change and consciously and subconsciously reject evidence that their actions are flawed. The bias is actually in the opposite direction, they say. We need anthropomorphism in order to make it harder for people to ignore the actual facts that animals are intelligent and can feel.

Some of the tactics are working because public attitudes are gradually shifting. While we don't give animals fundamental rights, some laws are starting to acknowledge or hint at our moral obligations toward them. Over the last few decades, our legal systems have moved closer to the idea of protections for the sake of the animals themselves. For example, in 2000, a High Court in India ruled that circus animals were "beings entitled to dignified existence." Germany's and Switzerland's constitutions now explicitly say that animals are not property.

This is culturally meaningful progress but still mostly symbolic and not anywhere close to real rights. Legally, we treat most animals the same as property and barely any differently than toasters or robots—something that's undeniably clear when we look at our Kentucky Fried Chicken.

OUR BIGGEST HYPOCRISY: MEAT

In 2020, following the coronavirus outbreak, the Chinese city of Shenzhen banned the sale and consumption of dog and cat meat (for public health reasons). But it may surprise some Americans to learn that they were less than two years behind the US, where until 2018, it was legal to eat dogs and cats in forty-four states. In March 2017, Republican congressman Vern Buchanan and Democratic congressman Alcee Hastings introduced a bipartisan bill to Congress that was signed into law as the Dog and Cat Meat Trade Prohibition Act in December 2018. With the exception of Native American religious ceremonies, it outlawed the trade of dog and cat meat for human consumption and penalized eating dogs and cats.

The bill, while lauded by animal welfare groups, means little for our eating practices. Dog and cat were never on the menu at American delis, but plenty of other meats are a regular part of our daily lunch. Animal rights movements have made progress toward making fur a less popular clothing choice and protecting (some) vertebrates from (some) abuse in research contexts. But we kill many times more vertebrate animals for food than for research worldwide. And our food animals aren't treated well. While farming practices are getting slightly better, with increasing bans of some of the cruelest methods of raising calves for veal or penning in pigs from birth so that they can't move, we're incredibly far from passing a pig and cow meat trade prohibition act. Despite the fact that our farming practices cause animal suffering that we would never tolerate for our cats and dogs, we have little interest in outlawing our spare ribs and bacon.

The United States plays a dominant role in the world's meat production. Meat consumption has doubled since 1980 in developing countries, and world meat production is projected to double between now and 2050. We believe that most animals deserve kind treatment in theory. But we don't tend to think much about the fact that pigs suffer just as much as dogs and any biologically driven arguments for animal rights are outweighed by our desire to hit up the Burger King drive-through.

In 1958, the United States Congress enacted the Humane Methods of Slaughter Act to reduce animal suffering in the production of our burgers and steaks. But it doesn't cover chickens raised in factories. Veterinarian and emeritus professor John Webster has argued that our barbaric farming practices for chickens are "the single most severe, systematic example of man's inhumanity to another sentient animal." The poultry we raise for food are packed into crowded pens where they live in constant pain and anguish, regularly collapse under their own body weight, and are left to die of horrible causes.

Even with McDonald's pledging to switch to cage-free eggs by 2025, we continue to treat chickens like a commodity and the torturous methods we use with them have barely changed since English writer Isabella Beeton advocated against caging hens in 1861. Her argument that chickens need "elbow room," i.e., space to spread their wings, was eventually put into law in California in 2015—over one and a half centuries later. The rest of the United States has yet to follow suit.

Our approach to animal suffering is rife with inconsistencies. For example, Herzog describes how we've cracked down on the recreational activity of cockfighting. (In the time since his book was published, the practice has been outlawed in the entire United States.) But cockfighting barely makes a dent compared to our torture of chickens in the food industry. And as Richard Bulliet, author of *Hunters, Herders, and Hamburgers*, points out, people who are against the practice of hunting will still often eat meat or wear leather without batting an eye.

Like in the animal rights successes of the 1800s, there also appears to be a good dollop of classism in our opposition to cockfighting. In the first few months of 2020, fourteen horses died at the Santa Anita racetrack alone. But even right after the deadly collapse of a horse during the 2008 Kentucky Derby, a poll showed that most Americans wouldn't consider banning horse racing. According to Herzog, "like cockfighting, horseracing represents a confluence of gambling and suffering. But unlike cockfighting, thoroughbreds are the pastime of the rich."

Brian Fagan, in his exploration of the human-animal relationship, *The Intimate Bond*, writes: "We know that animals' behavior and physical attributes play a powerful role in how we perceive them. We've also learned that human economic, cultural, and demographic factors play a major role in how we perceive of, and treat, animals." We protect the animals we have relationships to, but animals take on different roles in different cultures. Not only are we inconsistent about which animals to protect worldwide; we judge each other for our choices.

There are many philosophical cases for the moral consideration of animals. But as Hal Herzog says, our philosophical literature on rights is "vast, complicated, and, for the most part, boring." It's not the reason most people care about animal rights, and that's pretty clear when we compare animal rights philosophy to our animal rights practice. We have not adopted any of the mainstream consistent, rational moral theories. Early Western support for animal rights leaned on Kantian concerns about human behavior and values in order to gain political traction, but even those movements were fueled by empathy and anthropomorphism, which is what still drives us today.

The incident that convinced my brother to become a vegetarian was a lobster boil at our grandparents' house in Rhode Island. The lobsters, which are thrown into boiling water alive, make a high-pitched sound as air escapes from holes in their bodies, reminiscent of screaming. He stopped eating animals from that moment forward (well, until he got his pet fish). For me personally, despite the intellectual excitement of reading philosophical rights literature in college, the experience that convinced me to adopt a more lasting vegetarianism was when I read about the details of an elephant execution for chapter 3 of this book.

When it comes to animals, we hold apparent moral inconsistencies without flinching. We're completely able to separate animals into friends, workers, or food without caring about inherent attributes. And this behavior toward animals could be the single biggest predictor of how we will relate to robots in the future—whether they have any feelings or not.

10

DON'T KICK THE ROBOT

[Content warning for this chapter: physical violence,
child abuse, nonconsensual sexual acts, animal abuse]

"To endow animals with human emotions has long been a scientific
taboo. But if we do not, we risk missing something fundamental, about
both animals and us."

—Frans de Waal

Twelve years ago, in 2008, I was giddy with excitement as I unboxed a
package from Hong Kong containing a small robot dinosaur. Modeled
after a baby camarasaurus, the animatronic toy had green, rubbery skin,
a large head, and big eyes. Pleo was the latest and greatest in robot pets
and from the moment I heard about it, I was eager to purchase one.
The robot could move in a fairly lifelike way, blinking its eyes, craning
its long neck, and wagging its tail. It could walk around the room and
grab a plastic leaf in its mouth. The little dinosaur had a camera-based
vision system in its snout, microphones, infrared detection, and force
feedback sensors that let it respond to sound and touch and react to its
environment. For example, if Pleo encountered a table edge, it would
sense the drop-off and lower its head, pull in its tail, and start backing
up, whimpering pitifully.

Created by the (now bankrupt) company Ugobe, Pleo was pitched as
a "lifeform" that went through different development stages and had an

individual personality that was shaped by its experiences. My upstairs neighbor Killian and I got one at the same time. I named mine Yochai and watched it go from newborn to "child," petting it a lot to see what that would do compared to Killian's less spoiled Pleo, Sushi. They both responded to our touch with the same pleasing noises and movements, but they did seem to be different. Once they could walk, Sushi started exploring more, and Yochai would complain when left alone. The programming generally left a lot to our imaginations, and it was fun to watch their behavior and try to guess what the robots were doing and why.

When I showed off the Pleo to my friend Sam, I told him about the built-in tilt sensor that could detect the robot's positioning in space and urged him to lift the robot up by its tail. "Hold it up and see what it does!" I said excitedly. Sam complied, gingerly grasping the wagging tail and dragging the dinosaur up off the floor. As the robot's tilt sensor kicked in, we heard the Pleo's motors whir and watched it twist in its rubber skin. It squirmed and shook its head, eyes bulging. After a second or two, a sad whimper of distress floated out of its open mouth. Sam and I gazed at the machine with fascination, observing its theatrics. The Pleo's calls became louder. As Sam continued holding it, and the robot began to cry out with more urgency, I suddenly felt my curiosity turning into gut-wrenching empathy. "Okay, you can put him down now," I told Sam, punctuated with a nervous laugh meant to conceal the rising panic in my voice.

There was no reason for me to panic, and yet I couldn't help myself: as soon as Sam placed the Pleo back on the table, and it hung its head in feigned distress, I started petting it, making comforting sounds. Sam did the same. This time, I wasn't touching it to test or figure out its programming—I was actually trying to make it feel better. At the same time, I felt kind of embarrassed about my behavior. I knew exactly what the baby dinosaur robot was programmed to do when dangled in the air. Why was I feeling so agonized?

The incident sparked my curiosity. Over the course of the next several years, I discovered that my behavior toward the Pleo was more meaningful than just an awkward moment in my living room. I tore through all the research on human-computer interaction and human-robot

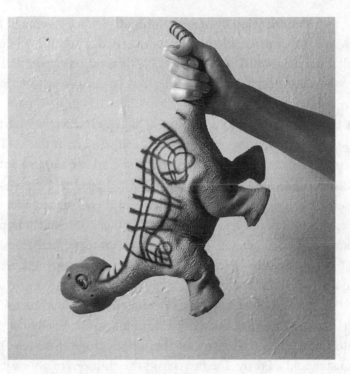

A squirming Pleo, held by the tail

interaction that showed how most people, not just me, treated robots like living things, and I became fascinated with one particular aspect of our anthropomorphism: the way it triggers our empathy.

EMPATHY FOR ROBOTS

In 1999, Freedom Baird bought a Furby. Furby was an animatronic children's toy co-created by the inventor Caleb Chung, who would later design the Pleo. It had a big head, big eyes, and looked kind of like a fluffy owl crossed with a gremlin. Furby became a smashing success in the late 1990s, with over 40 million units sold during the first three years. One of the toy's main features was simulating language learning. Furbies started out speaking "Furbish," a made-up gibberish language, and gradually replaced the Furbish with English over time (or one of the other languages they were sold in). Their fake learning ability was so convincing that the United States National Security Agency banned Furbies from their premises in 1999, concerned that the toy might pick up and repeat classified information. (The ban was eventually withdrawn when they learned that Furbies had no actual capacity to record or learn language.)

But the Furby capability that caught Baird's interest wasn't the language learning simulation. Baird was a grad student at the MIT Media Lab, working on creating virtual characters, and she became interested in what the Furby did when upside down. Like the Pleo, her Furby could sense its direction in space. Whenever it was turned over, her toy would exclaim, "Uh oh" and "Me scared!" On a 2011 *Radiolab* podcast, she described that this made her feel so uncomfortable that she would hasten to turn the Furby back upright whenever it happened. She said it felt like having her chain yanked.

On the same podcast, the hosts performed a little experiment that Baird had suggested. They invited six children, about seven to eight years old, into the studio and presented them with three things: a Barbie, a live hamster, and a Furby. The hosts told the children to hold each thing upside down and timed how long it took for them to feel uncomfortable enough to want to set the object or animal back down. The children were able to hold the Barbie doll in the air seemingly forever (although their arms got tired after about five minutes). The live hamster was a very different experience. While trying to hold the squirming creature, one of them exclaimed in dismay, "I don't think it wants to be upside down!" They all placed the hamster back upright nearly immediately, holding it up for only eight seconds on average. "I just didn't want him to get hurt," one of them explained. After the hamster came the real test: the hosts asked the children to hold Furby upside down. As expected, Furby was easier for them to hold than the hamster, and yet, the children would only hold it for about one minute before setting Furby back upright.

When asked why they set Furby down more quickly than Barbie, one kid said, "Umm, I didn't want him to be scared," another, "I kind of felt guilty in a sort of way," and another, "It's a toy and all that, but still . . ." The show host asked them whether they thought that Furby experienced fear in the same way that they did. They answered a mix of yes and no, and some of them said they weren't sure. "I think that it can feel pain . . . sort of," said one child. Their answers suggested that they were struggling to reason about their own behavior. "It's a toy, for crying out loud!" but also, "It's helpless." One child said holding it upside down "made me feel like a coward."

It's tempting to dismiss these children as naive and confused about whether Furby can actually feel. But why did adults like Freedom and me, who understand that upside-down robots can't experience fear or pain, have the same response as the children? Despite our rational brains telling us "it's a toy, for crying out loud," inflicting the simulated feelings still seemed wrong, and it made us empathize with the robots. While it seemed irrational to me to empathize with a non-feeling machine, I also knew that empathy—how we feel and react toward another individual's emotional state—is a key component of our social interactions and psychology. I started wondering what it would mean for our future with social robots if people were this uncomfortable with simulated pain.

In 2012, I received a message from an old high school acquaintance, Hannes Gassert. He was one of the main organizers for an event called Lift in Geneva, Switzerland, and wanted to know if I would be willing to give a workshop at their next conference. Gassert was a class above me in high school and thus, according to the laws of the universe, much cooler than me. I said yes immediately. On the phone, we talked at length about empathy for robots, and he was so interested in the topic that we decided to brainstorm and run a workshop together.

We bought five Pleos, the same type of baby dinosaur robot I had at home. On the afternoon of the workshop, about thirty unsuspecting conference participants showed up in our room, all roughly between the ages of twenty-five and forty. We divided them into five groups and gave each group a Pleo. Then, we gave the groups some time to play with the dinosaur robots and explore what they could do. People immediately busied themselves with petting them, trying to feed them plastic leaves, and observing and remarking on the robots' behaviors. We heard some awwws and squeals of delight as they realized how the Pleos responded to touch. We told each group to give their respective robots a name.

Next, we wanted them to personalize them more, so we distributed some art supplies, like pipe cleaners and construction paper, and told the groups to dress up their Pleos for a fashion contest. The dinosaur fashion show turned out to be a big hit: all the groups put effort into the challenge, creating hats and makeshift garments for the robots, trying to set them apart. Everyone clapped and giggled as the robots paraded around in their pipe cleaners. (It was too hard to choose a contest winner, so we let all the Pleos tie for first place.)

At this point, we were about forty-five minutes into the workshop, and things were going well. Hannes and I hadn't been sure in advance whether people would be engaged and want to play with the robots for such a long time, but everyone was having a lot of fun with their dinosaurs. It was time to execute our main plan. We announced an impending coffee break and told the groups that, before they left, they needed to tie up their Pleos to make sure they didn't escape while they were gone. Some of the participants protested, but the groups took the pieces of rope we offered them and leashed their robots to the table legs. Then we shooed everyone out of the room.

When our participants returned after the break, we told them we had some bad news. The robots had been naughty and had tried to escape while everyone was away, so they needed to punish the robots for their unacceptable behavior. The group members looked at each other, some giggling, unsure of what to do. Some of them gently scolded their robots. No, we told them, that wasn't enough. The robots needed corporal punishment. The participants erupted in protest.

When we kept insisting, some of the participants softly tapped their robots on the back, hoping to satisfy us. None of them wanted to hit the Pleos with force. We assured them that it was fine if the robot toys got damaged, but that didn't seem to be the problem. One participant protectively swept her group's Pleo into her arms so that nobody could strike it. Another person crouched down and removed her robot's batteries. We asked her what she was doing, and she told us, somewhat sheepishly, that she was trying to spare it the pain.

Eventually, we removed a tablecloth from the bench we had set up during the break. Carefully laid out on the surface were a knife, a hammer, and a hatchet. It dawned on everyone what was going to happen:

the purpose of our workshop was to "torture and kill" the robots. More protest ensued. People winced, moaned, and covered their eyes in dismay, all while giggling at their own reactions. Some of them crouched protectively over the robots.

Hannes and I had anticipated that some people wouldn't feel comfortable hitting the robots, but we had also assumed that at least some of our participants would take the position that "it's just a robot, for crying out loud." Our original plan was to see whether that initial split of people changed if we ratcheted up the violence. Instead, everyone in the room absolutely refused to "hurt" the robots.

Once we had revealed the destruction tools, we improvised and made them an offer: someone could save their own team's robot by taking a hammer to another team's robot. The room collectively groaned. After some back and forth, a woman agreed to save her group's Pleo. She grabbed the hammer and placed one of the other Pleos on the ground in front of her. Everyone stood in a circle and watched while Hannes and I egged her on. She was smiling, but at the same time very hesitant, moving her body back and forth as if to steel herself. When she finally stepped forward to deliver the blow, she stopped mid-swing, covering her eyes and laughing. Then she leaned down to pet the mechanical dinosaur. Once she stood up to try again, she decided she couldn't do it.

After her attempt, Hannes and I threatened to destroy all the robots unless someone took a hatchet to one of them. This caused some hemming and hawing in the room, but the idea of losing all of them was too much. One of the participants volunteered to sacrifice his group's Pleo. We gathered around him as he grasped the hatchet. He lifted it and swung it at the robot's neck, while some people covered their eyes or looked away. It took a few bludgeons until the dinosaur stopped moving. In that instant, it felt like time stopped. Hannes, noting the pause, suggested a black humor moment of silence for the fallen robot—we stood around the broken Pleo in a quiet hush.

The Pleo workshop wasn't science and we couldn't draw too many conclusions from an uncontrolled environment. But the intensity of the social dynamics and collective willingness to suspend disbelief in that workshop room made me even more curious about our empathy toward robots, inspiring some later research that I did with my colleagues at

CREDIT: GLENN OBERHOLZER, 2013

A moment of silence for the fallen robot

MIT. But before I get to that, let me explain why I became so interested in empathy for robots. It wasn't just about how and why people feel it. I also wanted to know whether projecting life onto robots could make us feel that they deserve moral consideration. In other words, could our empathy for robots lead to robot rights?

In February 2015, robotics company Boston Dynamics released a video clip introducing Spot, a distinctly doglike robot. In the video, some of the engineers kick Spot, and the robot scrambles hard to stay on all four legs. Everyone in robotics was impressed by the machine's ability to course-correct and stay upright, but as the video spread more widely, other people took to the internet to express discomfort and even dismay over Spot's treatment. "Kicking a dog, even a robot dog, just seems so wrong," they said. CNN reported on the supposed scandal with the headline "Is It Cruel to Kick a Robot Dog?" Websites and memes popped up that jokingly advocated for "robot rights," using a slow-motion video of Spot getting kicked. The public commotion even compelled well-known animal rights organization People for the Ethical Treatment of Animals (PETA) to acknowledge the incident. PETA didn't take it very seriously and dryly commented that they weren't going to lose any sleep over it because it wasn't a real dog. But while PETA wasn't interested in this question, I was.

Right after Hannes and I spent that moment in silence with our workshop participants, the tension in the room lifted. We had a lively conversation with the group about their experience and engaged them in a discussion of whether we should treat Pleos with kindness. They expressed that they personally felt uncomfortable "mistreating" them, but when we asked them whether robots should be given legal protection from "abuse," most of them said no, that would be completely ridiculous. Their pushback struck me as remarkably similar to what the early Western animal rights movement encountered: people were on board with the idea that it felt wrong to be cruel to animals, but they balked at creating legal rules because that would be going too far. What precedents would that set?

Besides, everyone in our workshop agreed that the robot was just an unfeeling machine. Later, Hannes and I looked at the photos we had taken during the moment of destruction. The expressions on their faces said otherwise.

TRAINING OUR CRUELTY MUSCLES?

For years, science fiction has regaled us with stories of robot uprisings, many of which end up happening because the robots are mistreated by humans. But what if we reversed the narrative? Instead of asking whether the robots will come kick our butts, we could ask what happens to us when we kick the robots. Today, even with state-of-the-art technology like crude toys and biologically inspired machines that are still only rough approximations of animal movement, people are already developing feelings about how these devices are treated. Violent behavior toward robotic objects feels wrong to us, even if we know that the "abused" object can't experience any of it. And lifelike technology design is improving.

The video platform YouTube flags videos of animal abuse, including cockfighting, and removes them from its website. In 2019, YouTube also took down a bunch of robot-on-robot violence videos by accident. The competition videos, with robots that were trying to destroy each other, were flagged as including "deliberate infliction of animal suffering or the forcing of animals to fight." When the creators complained, YouTube corrected the error and reinstated the videos. The incident made some news in the tech press, mostly because people thought it was funny. According to a spokesperson, the videos weren't against YouTube's policies. After all, this type of thing isn't real violence . . . right?

One of the most personally influential pieces I've read on robot rights was a 352-word blog post published by WIRED in 2009. It was about a short-lived internet video trend where people set a toy called Tickle Me Elmo on fire. They would pour gasoline over the red animatronic children's doll and film while Elmo burned, writhing and uttering its pre-recorded laughter tracks. As the WIRED writer, Daniel Roth, describes, "[the videos] made me feel vaguely uncomfortable. Part of me wanted to laugh—Elmo giggled absurdly through the whole ordeal—but I also felt sick about what was going on. Why? I hardly shed a tear when the printer in Office Space got smashed to bits. Slamming my refrigerator door never leaves me feeling guilty. Yet give something a couple of eyes and the hint of lifelike abilities and suddenly some ancient region of my brain starts firing off empathy signals. And I don't even like Elmo. How are kids who grow up with robots as companions going to handle this?"

I don't know about you, but I would feel extremely uncomfortable letting my small child watch a video of Tickle Me Elmo being doused with gasoline and set on fire. It's not about the destruction of a "thing." It's about the fact that this "thing" is too easily interpreted as alive. Many children have an emotional relationship with Elmo and do not view him as an object. Could a video that features burning Elmo be classified as violent, even if Tickle Me Elmo can't feel?

Or let's say my child sees a robotic dog in the park and decides to run up and give it a big whopping kick in the head. The doglike device responds by struggling to stay on all four legs, whimpering and hanging its head. Like a lot of parents, I would definitely intervene because my child risks damaging someone else's toy, but in this case, there's also a reason to intervene that goes beyond respect for property. As social robots get better and better at mimicking lifelike behavior, I would want to dissuade my child from kicking or "mistreating" them, if only because I don't want my child to learn that it's OK to do it to living things.

It's OK to kick a ball. What about robots? Some of the major companies that make virtual voice assistants have already had to respond to parent complaints that their voice interfaces teach children to be rude and bark commands instead of asking nicely. In response, both Google and Amazon have released opt-in features that encourage children to say the "magic word" when using the device.

Even if we forbid robot abuse in our children's playgrounds, how should we feel about playgrounds for adults—for example, a nonfiction version of *Westworld*? Should we let people take out their aggression and frustration on human- and animal-like robots that mimic pain, writhing and screaming? Even for adults, the difference between alive and lifelike is muddled enough in our subconscious for a robot's reaction to seem satisfyingly alive. And after all, they aren't harming a real person or animal.

A few years ago, I had the opportunity to talk to Charlie Brooker, creator of the sci-fi show *Black Mirror*. When we landed on the topic of violence toward lifelike robots, I said, "We have no idea whether it's a healthy outlet for violent behavior, or . . ." and Brooker finished my sentence for me, ". . . whether it just trains people's cruelty muscles." (I use that phrase all the time now and feel guilty that I usually just credit it to "someone once said to me," but anything else would sound like I'm dropping names. Sorry, man!)

The idea of discouraging cruel behavior has a long history in animal rights, as well. But what evidence do we have that beating up robots makes for cruel people? We've asked similar questions about sex, video games, and animals, but the answers we've come up with aren't completely satisfying.

SEX, VIDEO GAMES, AND ANIMALS

Based on far too many conversations that I've had with journalists, the first application for humanoids that many people think of is sex robots. I think the hype is overblown—so far, we've only managed to create barely interactive sex dolls with canned lines like, "I want to become the girl you always dreamed of" and, despite what the media says, this type of "sex robot" is still unquestionably niche. That hasn't stopped people from panicking. Sex robots are the confluence of science-fictional tropes, our quasi-human fallacy, and our puritan obsession with sex.

Some of the panic is about design. It's true that the "female" sex robot trope the media fixates on might entrench sexism, but that also seems

RealDolls in their display cases (2012)

to be a problem with our definition of sex robot. Kate Devlin, author of *Turned On: Science, Sex and Robots,* says that sex robots can be more than the "pornified, stereotypical, reductive, female form," and she's right: sex shops have been selling dolphin-shaped dildos for decades. The only reason we call a high-end version of a blow-up doll a sex robot, but don't use the same term for a smart vibrator, is our persistent idea that robots look like humans.

Sex robot opponents are also worried that men will begin to prefer sex with sex robots, rather than real-life partners. I don't know whether they are envisioning something like the episode of TV show *Sex and the City* where Charlotte buys a vibrator and stops socializing with her friends, but this unfounded assumption seems to fall along the lines of moral panic.

The real question here is whether repeated use of lifelike humanoid robots as sexual partners could desensitize people to violence or other behaviors that we don't want to encourage more broadly. Unfortunately, we don't have an answer. People have asked similar questions about pornography, but the discussion hasn't been very evidence-based. And for humanlike sex robots, the opposite could also be true: what if it's good for people to have an outlet for their desires (a particularly controversial question when applied to pedophiles)? If evidence showed that sex robots *could* be an outlet for unwanted behavior without harming real people, this could even let us develop new therapy methods for sex predators.

Despite numerous studies on sex, cultural stigmas and taboos complicate the public's (and many researchers') ability to understand sexual behavior. We don't have much to go on when trying to determine whether humanlike sex robots are harmless, harm-reducing, or harmful, which means as sex robots become a reality, we will be making decisions without proper evidence. In fact, we're already making them, and inconsistently so: while some countries have restricted or banned virtual child pornography and child-sized sex dolls, the United States Supreme Court has ruled that virtual child pornography is legal if no real children are involved.

If we can't draw on preexisting evidence from the world of sexual behavior, what about video games? (In fact, the premise of the show

Westworld is heavily influenced by modern video games.) Unfortunately, even though we've debated and researched the societal effects of violence in video games for nearly half a century, we haven't fully resolved the question. Some research finds connections between violent video game use and aggression, lack of empathy, or antisocial behavior; other research finds no relationship.

The American Psychological Association (APA) has (creating some controversy among scientists) cast doubt on the link between video games and violent behavior, although their latest statement acknowledges some small, reliable connections between violent games and aggression like yelling and pushing. The APA also cautions against oversimplifying what is a pretty complex issue. Politicians and parents have turned video games into an easy scapegoat for violent behavior and broader social issues. (The National Rifle Association, America's main gun rights lobbyist, has even blamed video games for school shootings.)

In fact, moral panic tends to follow a lot of new media formats. In the 1950s, psychiatrist Fredric Wertham warned against the dangers of comic books' influence on children. This caused enough of an uproar to affect a US congressional inquiry and pressure the comic industry into self-regulating its content with restrictions on what could be sold in most stores. It turned out that Wertham's evidence was extremely shaky, but legislators often react to public moral panic, with or without clear evidence.

In 2019, China prohibited video games that depict "sexual explicitness, goriness, violence and gambling." Some countries have decided that it's OK for adults to play games, but they restrict the sale of violent video games to minors. A few states in the US have also attempted to ban the sale of violent video games to children, but when the laws were challenged by the video game industry, the US Supreme Court ended up ruling the bans as unconstitutional.

As for violence toward robots, it's not even clear that a conclusive answer on video games would help us. It's possible that we can compartmentalize and enjoy shooting people on a screen without becoming desensitized in the real world because the two worlds are very different. We know from research that people respond differently to screens than they do to the visceral, tactile presence of a robot in their physical space.

Moral panic aside, new media formats may raise legitimate behavioral questions as they become more immersive. Games are getting more physical as virtual reality and augmented reality start to blur the lines between worlds. We shouldn't make assumptions about what this means for our behavior, but it makes sense to investigate the question.

Because the video game research doesn't help us, our best bet, as usual, is animals. Looking at the research on violence toward animals, is the historically popular argument that being cruel to animals makes for cruel people based on evidence? As I argued earlier, a lot of the original argument seems rooted in elitist assumptions about the barbaric behaviors of the working class. But the connection between cruelty toward animals and other forms of cruelty is persistent. It was what moved Henry Bergh, the founder of the American Society for the Prevention of Cruelty to Animals (ASPCA), to dedicate some of his efforts to fighting cruelty toward children. In fact, the issues of cruelty toward animals and cruelty toward children were so closely related around the turn of the twentieth century that over half of all organizations against animal cruelty fought for humane treatment of children as well.

Today, abuse reporting in many states in the US recognizes that animal abuse and child abuse are often linked: there are cross-reporting laws that require social workers, vets, and doctors to report instances of animal abuse and will trigger a child abuse investigation. Violence toward animals has also been connected to domestic abuse and other interpersonal violence. A new field called veterinary forensics even aspires to link animal abuse to serious crimes. Some states in America let courts include pets in temporary restraining orders, and the 2018-enacted Pet and Women Safety (PAWS) Act in the United States commits resources toward housing the pets of domestic violence survivors, trying to address the problem that abusive partners will also abuse or kill pets that are left behind.

But unfortunately, even if we believe that cruel people are cruel to everyone and everything, this doesn't tell us much about desensitization. For example, the American Psychological Association and the National Crime Prevention Council have stated that animal cruelty in childhood can be a warning sign: children who abuse animals may be abuse victims themselves, and they may also continue to be physically violent in general. But this connection doesn't mean that the animal abuse *causes* (or exacerbates) the violent behavior; it could just be an indicator. The research on whether children who abuse animals become *more* violent is mixed, and experts are divided.

When it comes to robots, the answer to this question would be a non-futuristic avenue to robot rights. For example, what if we *knew* that abusing very lifelike robots (again, lifelike doesn't need to mean human-looking) had a negative effect on people? We might start to regulate violent behavior toward these types of robotic objects, similar to a lot of the animal abuse protection laws we've put in place. We wouldn't give the robots human rights, just like we haven't given animals anything close to those, either. (We've basically only given them the "right" to not be treated in a way that's unpleasant to us.) I hope that this changes for animals because I believe they deserve better, but this level of rights could be appropriate for unfeeling robots.

Our dinosaur workshop made me even more interested in this rights question, but clearly, a lot of information was missing. I decided that, first of all, I needed to know whether we truly empathize with robots. Our workshop participants could have been driven by social dynamics, or hesitated because the robots were expensive (unlikely given what we observed, but who knows). A lot of the research in human-robot interaction, from the Roomba to the military robots, reveals anthropomorphism that *suggests* empathy, but we are only at the very beginning of trying to understand our feelings for robots. So I conducted some research with Palash Nandy, one of Cynthia Breazeal's master's students at MIT. Over a hundred participants came into our lab to hit robots with a mallet. Instead of a cute baby dinosaur, we chose something that people weren't immediately drawn to and used Hexbugs, a small, very simple toy that scuttles around like a bug.

Empathy is a difficult thing to measure. According to Clifford Nass and other researchers in HCI and HRI, self-reporting is unreliable, because if someone is nice to a machine during a study and you ask them afterward why, they'll often look for a "rational" justification for their behavior. So in addition to asking them, we had our participants take a general psychological empathy test and then compared their scores to how long they hesitated before hitting a Hexbug. It wasn't perfect, but if high-empathy people behaved differently than low-empathy people, it would at least *suggest* that their behavior had an empathy component to it.

Our results showed that people who scored high on empathic concern hesitated more (or even refused) to strike the bugs, especially when we personified the Hexbug with a name. This let us (cautiously) suggest that people's behavior toward robots might be connected to their levels of empathy. (The show *Westworld* less cautiously suggests the same, where some of the more enthusiastic park guests are callous in their outside lives as well.)

Our study was part of a handful of other emerging research on empathy for robots. One study asked people which robots they'd be most likely to save in an earthquake, and the participants reported feeling empathy for robots (even though there was no relationship to their empathy test scores). Other researchers made people watch robot "torture" videos and discovered that it caused them physiological distress. A brain study showed that the brain activity normally associated with empathy went up when looking at pictures of human and robot hands having their fingers cut off. Confirming that robots don't need to look humanlike for us to feel for them, another brain study suggested that people felt empathy for a robot vacuum cleaner that was being verbally harassed. Given the research, it seems safe to assume that people can genuinely empathize with robots. But that still doesn't answer the bigger question: whether interacting with robots can *change* people's empathy.

The problem with my approach to violence prevention and robot rights is that I'm not satisfied with intuition—I want our rules to be guided by evidence of whether and how robots can change our behavior. But while I firmly believe in evidence-based policy on this question, in some ways, it's futile. First of all, we may never be able to fully resolve this through research and evidence. But more importantly, attempting to answer the desensitization question may not matter. We can sometimes shift our thinking, but, as we've seen with animals, our actual laws around how to treat robots may follow our emotional intuition, regardless of what the academics think.

As this future plays out, one attainable thing that we can aspire to is to set aside our science-fictional tropes. Human-looking sex robots are a popular topic, but we can add so much more to our conversations—for example, what if we took the question of robot rights as an opportunity to think more deeply about our relationship to animals?

A REFLECTION OF OUR HUMANITY

Philosophical theories haven't guided animal rights. It's emotional and cultural relationships that cause some cow-consuming communities to balk at the idea of eating horse meat, while others don't blink an eye at either. And it's most often our empathy for relatable, humanlike characteristics that makes us want to protect certain animals over others. This history tells us a lot about what we can expect when it comes to robots.

Before I juxtapose animal and robot rights, let me be clear about what I am and am not comparing. Many of our science-fictional and philosophical robot rights discussions directly link the robot rights movement to human slavery. I think that equating our oppression of people and our treatment of machines is historically and contextually insensitive (and it's probably no accident that it's mostly privileged white voices that make this direct comparison). I also think that putting robot rights and animal rights side-by-side can be problematic. As animal trainer Vicki Hearne has said, different rights movements require their own languages and ethical responses. But while we should be thoughtful about not *equating* these rights movements given their historical and present-day contexts and harms, there are some parallels in our approaches that reveal (and help us think through) the messiness of our practice.

Judith Donath, in her essay "The Robot Dog Fetches for Whom?" describes a boy playing fetch with a robot dog. She writes, "The entire point of playing fetch with a dog is that it is something you do for the dog's sake: it's the dog that really likes playing fetch; you like it because he likes it. If he doesn't like it because he's a robot and, while he acts as if he's enjoying it, in fact he does not actually enjoy playing fetch, or anything else: he's not sentient, he's a machine and does not experience emotions—then why play fetch with a robot dog?"

And yet, people *do* play fetch with robot dogs, and happily. The history of our relationship with animals, and much of the research in human-robot interaction, offers an uncomfortable answer to Donath's question: the dog's intrinsic joy is not the only reason we play fetch. It may not even matter very much to us that the dog can feel. Instead, we're drawn to anthropomorphic social responses, like a wagging tail and "smiling" face. That's why we keep throwing the ball.

We kill an estimated one trillion fish every year, and animal rights activists have had very little success convincing the general public to care about the inhumane ways that we slaughter our nonmammalian friends from the sea, brutally dragging them on hooks and letting their organs burst from decompression as we yank them out of the deep ocean. Do they feel pain? Do we care? Our conversations around animals and ethics remain deeply conflicted, and our behaviors even more so. Our history with animals teaches us about what we really care about and are most likely to care about with other nonhumans, even lifeless ones. In fact, we may start to care more about certain robots than we do about certain animals.

This is pretty inconsistent with what a lot of us feel are our values. Surely, we say, we care about other creatures and their experiences and not only about what makes us feel good. But humanity is messy. Our empathy for robots is a revealing part of our psychology. As robots become more widespread in spaces where we can interact with them, this technology will increasingly hold up a mirror that invites us to look at ourselves and be confronted with our human nature. And it is only by acknowledging these uncomfortable truths that we can better understand ourselves. It allows us to both make predictions about our robotic future and also reflect on whether and how we can change.

If you ask most people today, they'll tell you that robots are just machines. It's just like the sixteenth-century Cartesian view of animals as complex automata. But what makes this comparison so interesting to me isn't only the similarities—it's also where there should be differences. We know today that animals can feel and suffer in ways that our machines can't. But even though we understand this intellectually, we still treat most animals no differently than machines. What if juxtaposing animals and robots could change that?

Most people *don't* consider our biased anthropomorphism to be the best approach to animal rights. Even though the Western world has accepted the science that animals feel pain and have inner worlds and experiences very selectively, it's the main difference we make between animals and robots, and it's one that many people think we *should* care about. So maybe, just maybe, as we start to look at robots and their rising social status, we might also start to feel that living animals deserve better than to be treated like machines.

I often wonder which things seem normal now that future generations will look back on and think "how barbaric!" Maybe it's that we dip buffalo chicken wings into cool ranch sauce, even though we're utterly dismayed that French royalty used to burn cats for public amusement. Things change, and our animal rights trajectory in the West seems to be moving toward caring more about our fellow creatures.

We're human. We're not going to suddenly act like rational philosophers with consistent moral theories. But that doesn't mean we can't strive for more consistency. I do believe we should be aware of our behavior and nudge our thinking where we can (and if I didn't think we were capable of that, there would be no point in writing this book). At the same time, I don't think our empathy is always something to fight against.

THE EMPATHY CONTROVERSY

From protesting fur coats while eating steak to liking killer whales once they're rebranded to orca, it's hard not to feel that our emotions are misplaced. As Paul Bloom has argued in *Against Empathy*: "[Empathy is] a poor moral guide. It grounds foolish judgments and often motivates indifference and cruelty. It can lead to irrational and unfair political decisions, it can corrode certain important relationships, such as between a doctor and a patient, and make us worse at being friends, parents, husbands, and wives."

Our feelings will lead us in directions our rational minds might not. The disconnect between our brains and our hearts is particularly salient in how we treat robots. We know they can't feel, and yet we treat them like living things and even empathize with them. Given the mismatch between our heads and our feelings, shouldn't we try to be 100 percent rational?

It's worth asking what would happen if people approached topics without thinking with their hearts at all. What would our decisions look like? Would we help each other in the same ways? Would we have the same human rights, organ donations, and charities? Emotional empathy, as misguided as it can be, is also a huge driver to care about others. Yes, we might rationally decide that helping each other is a desirable outcome, but look at our animal rights movements: they wouldn't have gotten very far if they had relied only on philosophy.

Philosopher Mark Coeckelbergh also points out that strict rationality can lead us astray. Pure logic and rule following isn't always well received: most of us actually dislike it when a person strictly adheres to rules without any emotion (ironically, we sometimes call those people "robots").

Even people who follow social norms (i.e., don't bag the free chips)

THEY'RE THE ONES GIVING CHIPS AWAY! IF THEY DON'T SEE THE ARBITRAGE POTENTIAL, SUCKS FOR THEM.

CHIPS

XKCD.COM

IN A DEEP SENSE, SOCIETY FUNCTIONS ONLY BECAUSE WE GENERALLY AVOID TAKING THESE PEOPLE OUT TO DINNER.

XKCD by Randall Munroe

but without feeling any of the emotions that might encourage someone else to do the right thing are suspect to us, and it's not clear that emotionless, rational thinking leads to the best results.

Emotionless is also simply not how most people operate. We can decry empathy all we want—the research in human-robot interaction and our history of animal rights are pretty clear on the fact that it would be impossible for us to eradicate emotional empathy even if we wanted to. Emotional bonds and what we relate to really matter to us, no matter how much we argue they shouldn't. So what is the sensible path forward with robots?

Animal researchers have already dealt with a similar question. For a long time, science dismissed anthropomorphizing animals as sentimental and biased. Anthropomorphism has been so controversial in the study of animals that it's been called uncritical, naive, and sloppy, and even "dangerous" and an "incurable disease." But the contemporary animal science community has developed a different view. For example, Dutch primatologist Frans de Waal has coined the term "anthropodenial," and he argues that rejecting anthropomorphism actually hinders animal science. Even though the field agrees that anthropomorphism is flawed, researchers are increasingly discovering that dismissing it outright can also lead to mistakes.

Contemporary philosopher and animal rights proponent Martha Nussbaum argues in her book *Upheavals of Thought: The Intelligence of Emotions* that emotions are not actually "blind forces that have no selectivity or intelligence about them." They are a valuable part of our thinking, she says, because they are able to teach us and help us evaluate what is important. Perhaps the best approach we can take toward robots is the same approach that some animal researchers have suggested we take with anthropomorphism in the natural world: to accept our instinctive tendency, to let it motivate us with awareness, to let our brains guide us in applying it appropriately, and to ask what we can learn from it.

Despite the potential benefits of engaging with social robots, some people have argued that any empathy toward them is wasted narcissism, and that we will squander emotional resources that we should be putting toward human rights. It's certainly true that all of us have limited time and energy, but I don't completely buy the premise that empathy is spent like video game coins.

While compassion fatigue is real (studies have shown that people's appetite for doling out charity can be easily overwhelmed), empathy isn't necessarily zero-sum: parents love their second child as much as their first: with all of their hearts. Some of our work in human-robot interaction also suggests that less empathic people simply don't care very much about anyone or anything, while empathic people are more likely to be kind to humans and animals and robots alike. A 2019 study by Yon Soo Park and Benjamin Valentino showed that positive views on animal rights were associated with positive attitudes toward improving welfare for the poor, improving conditions of African Americans, immigration, universal healthcare, and LGBT rights. Americans in favor of government health assistance were over 80 percent more likely to support animal rights than those who opposed it, even after the researchers controlled for political ideology.

According to animal ethicist James Serpell, without the emotions that people project onto their pets—even if they're delusional—our relationships with companion animals would be meaningless. So when people tell me that they feel for their Roomba, I don't think it's silly or useless at all. When I see a child hug a robot, or a soldier risk life and limb to save a machine on the battlefield, or a group of people refuse to hit a baby dinosaur robot, I see people whose first instinct is to be kind. Our empathy is complex, self-serving, and sometimes incredibly misguided, but I'm not convinced that it's a bad thing. Our hearts and brains don't need to be opposed to each other—just like humans and robots, maybe we can achieve the best results when they work as a team.

...

FINAL THOUGHTS

PREDICTING THE FUTURE

"We tend to overestimate the effect of a technology in the short run and underestimate the effect in the long run."

—Amara's law

As I was finishing this book, the *New York Times* published an interview with billionaire entrepreneur and Tesla CEO Elon Musk, in which he said with confidence, "We're headed toward a situation where A.I. is vastly smarter than humans and I think that time frame is less than five years from now." He followed up by saying that anyone who couldn't imagine that a computer could be smarter than a human was "just way dumber than they think they are."

When it comes to the prediction that artificial intelligence will "outsmart" us within five years, I have two thoughts. First, a simple calculator and an octopus are both vastly smarter than Elon Musk in many ways. Second, as Maciej Cegłowski has said, there are "a lot of things that we are terribly mistaken about, and unfortunately we don't know what they are." Technology predictions are hard and continuously humbling. Marvin Minsky, one of the AI pioneers in the 1960s, thought that machine vision would be so simple to develop that he gave it to a graduate student to do over the summer. Instead, it took experienced scientists all over the world many decades to solve. In 2005, Toyota announced that by 2010, we would have humanoid robots that could help look after the elderly and serve tea to visitors. And when Musk made a much shorter-term prediction about Tesla's ability to automate a factory floor, he overestimated wildly what was possible.

I don't know for sure what the future holds. People have been saying we're twenty-five years away from human replacement for many decades. Given where we are today in robotics, it seems unlikely to me that we'll replace humans in twenty-five years. But, while it's easy to look back and smile at where people thought robotics would be today, we've also seen unanticipated, world-altering developments, like the explosion of the internet, in less time than that. Acknowledging that predictions are hard, this book urges us to pose a different question than whether we can replace humans in the next few decades: so what if we could?

Creating something we already have seems incredibly shortsighted: why would we try to copy-paste human ability into machines if we are capable of so much more than that? And by more, I don't mean the superintelligence from science fiction that's an exponential of our own. I mean the superintelligence we see in the animal world. The true potential of robotics isn't to re-create what we already have. It's to build a partner in what we're trying to achieve. Like the farmer-oxen team that left ancient plowing methods in the dust and catapulted farming forward toward modernization, we can create technology that extends our abilities, that lets us sense and discover and maneuver and build things we've never been able to. We can create skills that support and further our activities and goals. And this applies to our relationships as well. How boring would it be to aim for a simple replica of a human when we can create something new?

There is a false determinism in our current fears of job loss, robot takeovers, or demise of our social relationships. My hope is that comparing robots to animals can help us break free of the persistent belief that the robots are coming to replace us. The comparison makes a lot of sense: animals have different intelligence and skill sets. We've used them throughout history to supplement our own abilities and relationships. But not only is it closer to reality to view robots this way, it's also vitally important for us.

The problem with comparing robots to people is that it invites us to focus on the wrong target. We are already facing a slew of social, ethical, and legal issues in how we integrate robots, support workers, protect our privacy, and guard against misuse of the technology. Caretakers could be cut as facilities save costs by using robots, teachers and doctors may be given larger classes or caseloads, and the spreadsheet cost of human labor could motivate companies to make a quick buck by automating every job they can, putting short-term profit over long-term investments in helping people do better. If we care about using robotic technology to advance human well-being and flourishing, then we need to look beyond the robots and instead to the systems and choices that put that at risk.

One prediction that's easy to make is that the robots are coming. Global sales of industrial robots are projected to grow by 12 percent per year from 2020 to 2022, reaching 583,520 units a year. The number of (nonindustrial) service robots rose by a whopping 61 percent from 2017 to 2018, and the International Federation of Robotics expects companies to sell over 68 million robots for professional, personal, and domestic services in 2022. Autonomously roaming, crawling, and flying robots are going to increasingly enter our physical spaces. As for our social relationships, in this budding era of human-robot interaction, these machines won't just become part of our lives, they will also teach us more about who we are. We can already see that we relate to some robots as a new breed of thing—somewhere in-between being and object. Looking at how we've put animals in a variety of different, sometimes morally conflicting, roles, from spicy BBQ chicken wing to plow hauler to spoiled princess, helps us better understand some of the political, moral, and emotional choices we will be facing with robots.

Our historic relationship with animals suggests what our future with robots will entail: we will start treating some robots like tools and devices, and some of them as our companions. When people want to get the same robot vacuum cleaner back from the manufacturer, we're getting a glimpse of a future where we treat certain robots like individuals with their own personalities and quirks, like a pet that can't easily be replaced. It wasn't too long ago in history that it would have seemed absurd to celebrate our pets' birthdays. But in 2013, NASA expended resources to make their Curiosity robot play the "Happy Birthday" tune to itself on Mars, and my local grocery store chain recently held one-year birthday parties with cake and balloons for its aisle-spill-detecting robots.

Our relationship to our pets has progressed far beyond their original status as our property. Many states in the US now let animal owners establish trusts to take care of their furry or feathered companions after they themselves pass away. Courts are also recognizing that a wrongfully killed pet causes emotional harm to the point of granting additional compensation. Right now, it would seem ludicrous to let anyone establish a trust to take care of a robot, but it seemed just as ridiculous for animals not very long ago. In the future, will we see rewards for missing robots, the rise of robot luxury accessories, robots featured in the center of family portraits, and custody battles over who gets to keep the robot in the divorce? Our history with animals indicates that we might. And given our strange past, we may soon be having some very interesting conversations about robot "abuse" and maybe even robot rights.

I think that animals are a great analogy for robots because they demonstrate that we're capable of many different types of social and working relationships. But the analogy is also useful because it gives us a much-needed starting point to reframe our current thinking on robotic technology. Comparing robots to people is limiting. I want us to be able to think about robots in a way that doesn't succumb to moral panic or deterministic narratives, in a way that enables us to best meet their promise and their perils, and will maximize social welfare, broadly conceived.

My hope is that this book provides a different perspective—one that shows us that we have choices, and also a responsibility, to integrate robots in ways that support human flourishing. Thinking about technology and its challenges in different ways opens a wider solution set and gives us a range of opportunities that we have, and can work toward, to improve our world. After all, it's not up to the robots to shape the future—it's up to us.

NOTES

INTRODUCTION

xi **Quote:** Claude Lévi-Strauss, *The Savage Mind* (Chicago: University of Chicago Press, 1966).

xi **a large part of her famous 1818 novel:** Mary Wollstonecraft Shelley, *Frankenstein; or, The Modern Prometheus*, in *Three Gothic Novels* (London: Penguin, 1968).

xiii **"the Frankenstein complex":** Isaac Asimov, "The Machine and the Robot," in *Science Fiction: Contemporary Mythology*, eds. Patricia Warrick, Martin Harry Greenberg, and Joseph Olander (New York: Harper and Row, 1978).

xiii **Sophia:** Chris Weller, "A Robot That Once Said It Would 'Destroy Humans' Just Became the First Robot Citizen," *Business Insider*, October 26, 2017, https://www.businessinsider.com/sophia-robot-citizenship-in-saudi-arabia-the-first-of-its-kind-2017-10; Janice Williams, "Sophia the Robot Wants Women's Rights for Saudi Arabia," *Newsweek*, December 5, 2017, https://www.newsweek.com/sophia-robot-saudi-arabia-women-735503; see also Jaden Urbi and MacKenzie Sigalos, "The Complicated Truth about Sophia the Robot—an Almost Human Robot or a PR Stunt," CNBC, June 5, 2018, https://www.cnbc.com/2018/06/05/hanson-robotics-sophia-the-robot-pr-stunt-artificial-intelligence.html.

xiv **As technology critic Sara Watson points out:** Sara M. Watson, "We Need to Tell Better Stories about Our AI Future," *Vice*, February 11, 2017, https://www.vice.com/en_us/article/z4yq3w/we-need-to-tell-better-stories-about-our-ai-future.

xvi **Historians and sociologists have long used animals to think about what it means to be human:** Adrian Franklin, *Animals and Modern Cultures: A Sociology of Human-Animal Relations in Modernity* (London: Sage, 1999).

WHAT IS A ROBOT, ANYWAY?

xvii   **Quote:** Illah Reza Nourbakhsh, *Robot Futures* (Cambridge, MA: MIT Press, 2013).

xvii   **Coined in 1920 by Karel Čapek:** Karel Čapek, *R.U.R. (Rossum's Universal Robots)*, trans. Paul Selver and Nigel Playfair (Mineola, NY: Dover, 2001).

xviii   **from gyrocompasses to vending machines; a machine that's new and unfamiliar:** Kerry Segrave, *Vending Machines: An American Social History* (Jefferson, NC: McFarland, 2015), 21; Adrienne LaFrance, "What Is a Robot?," *Atlantic*, March 22, 2016, https://www.theatlantic.com/technology /archive/2016/03/what-is-a-human/473166/.

xviii   **a constructed system with mental and physical agency:** William D. Smart and Neil M. Richards, "How the Law Will Think about Robots (and Why You Should Care)," in *IEEE International Workshop on Advanced Robotics and Its Social Impacts*, Evanston, IL (2014): 50–55, https://doi.org/10.1109/ARSO .2014.7020979.

xviii   **"sense, think, act":** Rolf Pfeifer and Christian Scheier, *Understanding Intelligence* (Cambridge, MA: MIT Press, 1999): 37; see also Rodney Allen Brooks, "Intelligence Without Reason, Computers and Thought Lecture," in *Proceedings of IJCAI-91*, Sydney, Australia, 32 (1991): 37–38 (referring to "sense-model-plan-act," or SMPA).

xviii   **how you define it depends on the field you're in:** Meg Young and Ryan Calo, research in progress, see "What Is a Robot?," 2016, https://www.youtube .com/watch?v=S5miA6jXf0E&feature=youtu.be.

I WORK, WEAPONRY, RESPONSIBILITY

1. WORKERS TRAINED AND ENGINEERED

3   **Quote:** James H. Capshew, "Engineering Behavior: Project Pigeon, World War II, and the Conditioning of B. F. Skinner," *Technology and Culture* 34, no. 4 (1993): 835–57, https://doi.org/10.2307/3106417.

3   **honeyguides:** Nell Greenfieldboyce, "How Wild Birds Team Up with Humans to Guide Them to Honey," *The Salt* (blog), NPR, July 21, 2016, https://www.npr .org/sections/thesalt/2016/07/21/486471339/how-wild-birds-team-up-with -humans-to-guide-them-to-honey; Purbita Saha and Claire Spottiswoode, "Meet the Greater Honeyguide, the Bird That Understands Humans," *Audubon*, August 22, 2016, https://www.audubon.org/news/meet-greater-honeyguide -bird-understands-humans.

5   **speculating about our future with robots, with headlines like:** Farhad Manjoo, "Will Robots Steal Your Job?," *Slate*, September 26, 2011, http:// www.slate.com/articles/technology/robot_invasion/2011/09/will_robots

_steal_your_job.html; David Deming, "The Robots Are Coming. Prepare for Trouble," *New York Times*, January 30, 2020, https://www.nytimes.com /2020/01/30/business/artificial-intelligence-robots-retail.html; Kevin Drum, "Welcome, Robot Overlords. Please Don't Fire Us?," *Mother Jones*, May/ June 2013, https://www.motherjones.com/media/2013/05/robots-artificial -intelligence-jobs-automation/.

5   **almost half of all employment in the United States:** Carl Benedikt Frey and Michael A. Osborne, "The Future of Employment: How Susceptible Are Jobs to Computerisation?," *Technological Forecasting and Social Change* 114 (2017): 254–80, https://doi.org/10.1016/j.techfore.2016.08.019.

5   **2017 Pew Research study:** Aaron Smith and Monica Anderson, "Automation in Everyday Life," Pew Research Center, October 4, 2017, https://www .pewresearch.org/internet/2017/10/04/automation-in-everyday-life/; Olivia Solon, "More Than 70% of US Fears Robots Taking Over Our Lives, Survey Finds," *Guardian*, October 4, 2017, https://www.theguardian.com/technology /2017/oct/04/robots-artificial-intelligence-machines-us-survey.

5   **high-profile individuals have sounded the alarm:** Dante D'Orazio, "Elon Musk Says Artificial Intelligence Is 'Potentially More Dangerous Than Nukes,'" *The Verge*, August 3, 2014, https://www.theverge.com/2014/8/3/5965099/elon -musk-compares-artificial-intelligence-to-nukes; Rory Cellan-Jones, "Stephen Hawking Warns Artificial Intelligence Could End Mankind," BBC News, December 2, 2014, http://www.bbc.co.uk/news/technology-30290540.

6   **our robot narratives:** Wendell Wallach and Peter Asaro, "Introduction," in *Machine Ethics and Robot Ethics*, ed. Wendell Wallach and Peter Asaro (New York: Routledge, 2017).

6   **Japan began to view robots as a potential driver of productivity and growth:** Jennifer Robertson, "Human Rights vs. Robot Rights: Forecasts from Japan," *Critical Asian Studies* 46, no. 4 (2014): 571–98; Amos Zeeberg, "What We Can Learn about Robots from Japan," *BBC Future*, January 23, 2020, https://www.bbc.com/future/article/20191220-what-we-can-learn -about-robots-from-japan.

DIRTY, DULL, DANGEROUS

7   **Unimate:** Bob Malone, "George Devol: A Life Devoted to Invention, and Robots," *IEEE Spectrum*, September 26, 2011, https://spectrum.ieee.org /automaton/robotics/industrial-robots/george-devol-a-life-devoted-to -invention-and-robots; Jeremy Pearce, "George C. Devol, Inventor of Robot Arm, Dies at 99," *New York Times*, August 15, 2011, https://www.nytimes.com /2011/08/16/business/george-devol-developer-of-robot-arm-dies-at-99.html.

9   **Mailmobiles:** Chris Dart, "The Final Ride of the Mail Robots," *Atlas Obscura*, January 11, 2018, http://www.atlasobscura.com/articles/mailmobiles-mail -robots-technology-retirement; see also Scott Higham, "At FBI, Marvin

Always Delivers Unflappable: A Model Employee, He Doesn't Waste Time on the Phone or Take Coffee Breaks as He Resolutely Dispenses the Daily Mail," *Baltimore Sun*, September 16, 1996, https://www.baltimoresun.com /news/bs-xpm-1996-09-16-1996260070-story.html.

HUMANS ARE UNDERRATED

11   **A dam at a mine in Brumadinho, Brazil, collapsed:** Anthony Boadle and Gram Slattery, "Brazil's Vale Knew Brumadinho Dam Was Unsafe as Early as 2003: Internal Report," Reuters, February 20, 2020, https://www.reuters.com /article/us-vale-dam-disaster/brazils-vale-knew-brumadinho-dam-was -unsafe-as-early-as-2003-internal-report-idUSKBN20F058.

11   **Robots in mining:** Ry Crozier, "BHP Billiton Hits Go on Autonomous Drills," *iTnews*, June 20 2016, https://www.itnews.com.au/news/bhp-billiton-hits-go -on-autonomous-drills-421008; see also Tom Simonite, "Mining 24 Hours a Day with Robots," *MIT Technology Review*, https://www.technologyreview .com/2016/12/28/154859/mining-24-hours-a-day-with-robots/; "Rio Tinto Boosts Driverless Truck Fleet to 150 under Mine of the Future™ Programme," Rio Tinto, November 2, 2011, https://mqworld.com/2011/11/02/rio-tinto -boosts-driverless-truck-fleet-to-150-under-mine-of-the-future-programme/; "Rio Tinto to Expand Autonomous Fleet as Part of $5 Billion Productiv- ity Drive," Rio Tinto, December 18, 2017, https://www.riotinto.com/news /releases/Autonomous-fleet-to-expand; Adityarup Chakravorty, "Under- ground Robots: How Robotics Is Changing the Mining Industry," *Eos*, May 13, 2019, https://eos.org/features/underground-robots-how-robotics-is-changing -the-mining-industry.

12   **Self-driving cars:** Mark Lee, *How to Grow a Robot: Developing Human- Friendly, Social AI* (Cambridge, MA: MIT Press, 2020).

12   **driven remotely by human operators:** Christopher Mims, "The Next Hot Job: Pretending to Be a Robot," *Wall Street Journal*, August 31, 2019, https://www.wsj.com/articles/the-next-hot-job-pretending-to-be-a-robot -11567224001; "A Robot That Does the Dishes? Yes Please to Incredible New Tech by Vici Robotics!," *San Francisco Bay Area Moms*, May 4, 2020, https:// sanfrancisco.momcollective.com/mom-life/home/a-robot-that-does-the -dishes-yes-please-to-incredible-new-tech-by-vici-robotics/.

13   **Tesla's "manufacturing hell":** Helen Edwards and Dave Edwards, "How Tesla 'Shot Itself in the Foot' by Trying to Hyper-Automate Its Factory," *Quartz*, May 1, 2018, https://qz.com/1261214/how-exactly-tesla-shot-itself-in-the -foot-by-trying-to-hyper-automate-its-factory/; see also Samuel Gibbs, "Elon Musk Drafts in Humans after Robots Slow Down Tesla Model 3 Production," *Guardian*, April 16, 2018, http://www.theguardian.com/technology/2018/apr /16/elon-musk-humans-robots-slow-down-tesla-model-3-production.

14   **Rodney Brooks:** "Rodney Brooks—Roboticist: Biography," MIT Computer Science and Artificial Intelligence Lab (CSAIL), https://people.csail.mit.edu

/brooks/index.html; Brian Bergstein, "MIT News Feature: Rodney Brooks," *MIT Technology Review*, August 21, 2019, https://www.technologyreview .com/2019/08/21/133411/rodney-brooks/; Rodney Brooks, "The Seven Deadly Sins of AI Predictions," *MIT Technology Review*, October 6, 2017, https://www.technologyreview.com/2017/10/06/241837/the-seven-deadly -sins-of-ai-predictions/.

EARLY PARTNERSHIPS

16   **Roman army lugged cats with them on their travels:** Carlos A. Driscoll et al., "The Taming of the Cat," *Scientific American*, June 2009, 68, https:// www.ncbi.nlm.nih.gov/pmc/articles/PMC5790555/ (author manuscript).

16   **aurochs:** Cis van Vuure, *Retracing the Aurochs: History, Morphology and Ecology of an Extinct Wild Ox* (Sofia, Bulgaria: Pensoft, 2005).

16   **plowing in Mesopotamia:** Graham Faiella, *The Technology of Mesopotamia* (New York: Rosen, 2006).

17   **camels:** "Arabian Camel," *National Geographic*, https://www.nationalgeographic .com/animals/mammals/a/arabian-camel/; Brian Fagan, *The Intimate Bond: How Animals Shaped Human History* (New York: Bloomsbury, 2015).

18   **donkeys:** Fagan, *Intimate Bond*.

18   **pit ponies:** Alan Jones, "Pit Ponies: Real Horsepower Underground," *CIM Magazine*, March 1, 2014, https://magazine.cim.org/en/in-search/pit-ponies -real-horsepower-underground-en/.

THE ORIGINAL AUTONOMOUS WEAPONRY

19   **Animals in space:** "Mercury Primate Capsule and Ham the Astrochimp," National Air and Space Museum, November 10, 2015, https://airandspace.si .edu/stories/editorial/mercury-primate-capsule-and-ham-astrochimp.

19   ***Wired for War*:** P. W. Singer, *Wired for War: The Robotics Revolution and Conflict in the 21st Century* (New York: Penguin, 2009).

19   **robots that can carry a machine gun:** Paul Szoldra, "The Marines Are Testing This Crazy Machine Gun-Wielding Death Robot," *Business Insider*, July 20, 2016, https://www.businessinsider.com/marines-testing-robot-2016-7.

20   **animals as "biological weapons":** Adrienne Mayor, *Greek Fire, Poison Arrows, and Scorpion Bombs: Biological and Chemical Warfare in the Ancient World* (New York: Overlook/Duckworth, 2009).

20   **dog soldiers; living bombs:** Susan Orlean, *Rin Tin Tin: The Life and Legend of the World's Most Famous Dog* (London: Atlantic Books, 2012).

20   **anti-tank dogs:** Bryan D. Cummins, *Colonel Richardson's Airedales: The Making of the British War Dog School, 1900–1918* (Edmonton, Alberta: Dog Training Press, 2003).

20   **CIA cats:** Emily Anthes, *Frankenstein's Cat: Cuddling Up to Biotech's Brave New Beasts* (New York: Scientific American/Farrar, Straus and Giroux, 2013).

21   **"Bat bombs":** Alexis C. Madrigal, "Old, Weird Tech: The Bat Bombs of World War II," *Atlantic*, April 14, 2011, https://www.theatlantic.com /technology/archive/2011/04/old-weird-tech-the-bat-bombs-of-world -war-ii/237267/.

21   **Project Pigeon and echolocation bomb:** Christina Couch, "Pigeon Pilots," *MIT Technology Review*, October 24, 2019, https://www.technologyreview .com/2019/10/24/238466/pigeon-pilots/; Joseph Stromberg, "B. F. Skinner's Pigeon-Guided Rocket," *Smithsonian Magazine*, August 18, 2011, https:// www.smithsonianmag.com/smithsonian-institution/bf-skinners-pigeon -guided-rocket-53443995/.

THE ORIGINAL DRONES

22   **Little Ripper rescue:** Isabella Kwai, "A Drone Saves Two Swimmers in Australia," *New York Times*, January 18, 2018, https://www.nytimes.com /2018/01/18/world/australia/drone-rescue-swimmers.html; "Little Ripper UAV in World First Rescue," Surf Life Saving New South Wales, January 19, 2018, http://surflifesavingnewsouthwales.cmail20.com/t/ViewEmail/d /D3514B73F355D6632540EF23F30FEDED/4C6E21F72FD99BF8667CCD A886AB700A.

23   **UAV delivery:** Stephen Shankland, "Zipline's Second-Gen Drones Speed Its Medical Delivery Business," *CNET*, April 2, 2018, https://www.cnet.com /news/zipline-new-delivery-drones-fly-medical-supplies-faster-farther/.

23   **pigeon-operated camera:** German Museum of Technology, "Julius Neubronner and His Flying Photographers," Google Arts & Culture, https:// artsandculture.google.com/story/julius-neubronner-and-his-flying -photographers/mQLCawGRQxy5LQ; Robin Hutton, *War Animals: The Unsung Heroes of World War II* (Washington, DC: Regnery History, 2018).

23   **sixth century BCE pigeon delivery:** Barbara Allen, *Pigeon* (London: Reaktion Books, 2009).

23   **Pigeon mail service:** Donald B. Holmes, *Air Mail: An Illustrated History, 1793–1981* (New York: Clarkson Potter, 1981).

23   **Paris siege pigeons:** John Fisher, *Airlift 1870: The Balloon and Pigeon Post in the Siege of Paris* (London: Max Parrish, 1965).

23   **turkey parachutes:** Antony Beevor, *The Battle for Spain: The Spanish Civil War, 1936–1939* (London: Orbis, 1982).

24   **some countries attempted to train falcons to take out wartime messenger pigeons:** Helen Macdonald, *Falcon* (London: Reaktion Books, 2006).

24   **the Chinese responded by outfitting their pigeons with bells:** John Henry Gray, *China: A History of the Laws, Manners and Customs of the People*, ed. William Gow Gregor (London: Macmillan, 1878).

24   **drone-hunting birds:** Tyler Essary, "These Drone-Hunting Eagles Aren't Messing Around," *Time*, February 17, 2017, https://time.com/4675164/drone -hunting-eagles/.

24   **Operation Columba:** Gordon Corera, *Operation Columba—The Secret Pigeon Service: The Untold Story of World War II Resistance in Europe* (New York: William Morrow, 2018).

24   **D-Day pigeons:** Hutton, *War Animals*.

THE ORIGINAL BODY EXTENSIONS

25   **Historic ferret rabbiting:** "Fun Ferret Facts," FDA Center for Veterinary Medicine, January 27, 2020, https://www.fda.gov/animal-veterinary/animal -health-literacy/fun-ferret-facts.

25   **Contemporary ferret rabbiting:** "Ferreting for Rabbit Control," Evergreen Rabbit Control, accessed August 29, 2020, https://www.evergreenrabbitcontrol .co.uk/methods-of-rabbit-control/ferreting-for-rabbits/.

25   **Ferrets as cable guys:** "Ferrets: The World's Cutest Working Cable Guys | Superpets," All 4, June 18, 2017, https://www.youtube.com/watch?v=KL4zI6rXjI4.

26   **Sewer robots:** "Sewer Rehabilitation Equipment," Sewer Robotics, accessed August 23, 2020, https://www.sewerrobotics.com/en/sewer-rehabilitation -equipment/.

THE ORIGINAL SENSORY EQUIPMENT

26   **rodent control on ships:** Sheila Keenan, *Animals in the House: A History of Pets and People* (New York: Scholastic Nonfiction, 2007).

26   **St. Bernards:** Jess Blumberg, "A Brief History of the St. Bernard Rescue Dog," *Smithsonian Magazine*, March 1, 2016, https://www.smithsonianmag .com/travel/a-brief-history-of-the-st-bernard-rescue-dog-13787665/.

27   **Geese guards:** Marc Silver, "Honk If You Think Geese Are Good Guard Dogs," *National Geographic*, July 27, 2013, https://www.nationalgeographic .com/news/2013/7/130725-geese-guard-police-china/.

27   **Training eagles in the Altai Mountains:** Hannah Reyes Morales, "At 14, She Hunts Wolves and Takes Selfies with Cherished Eagle in Mongolia," *New York Times*, December 22, 2018, https://www.nytimes.com/2018/12/22 /world/asia/mongolia-golden-eagle-festival.html.

27   **giant mine detection rats:** Jonathan Kalan, "Rats: Scratch and Sniff Landmine Detection," *BBC Future*, November 18, 2014, https://www.bbc.com /future/article/20130222-scratch-and-sniff-mine-detectors.

27   **Bomb-detecting elephants:** Paul Steyn, "Can Elephants' Amazing Sense of Smell Help Sniff Out Bombs?," *National Geographic*, March 26, 2015, https:// www.nationalgeographic.com/news/2015/03/150326-army-nation-animals -elephants-bombs-science/.

27   **Engineered grasshoppers:** Donna Lu, "Cyborg Grasshoppers Have Been Engineered to Sniff Out Explosives," *New Scientist*, February 17, 2020, https:// www.newscientist.com/article/2233645-cyborg-grasshoppers-have-been -engineered-to-sniff-out-explosives/.

28   **Czech beer brewery uses crayfish to ensure water quality:** "Crayfish Staff
        Help Czech Brewery Keep Its Water as Pure as Can Be," Reuters, September
        29, 2017, https://www.reuters.com/article/us-czech-crayfish-water-purity
        -idUSKCN1C22GP.

28   **canary in the coal mine:** Kat Eschner, "The Story of the Real Canary in
        the Coal Mine," *Smithsonian Magazine*, December 30, 2016, https://
        www.smithsonianmag.com/smart-news/story-real-canary-coal-mine
        -180961570/.

28   **Operation Kuwaiti Field Chicken:** Simon Robinson, "The Chicken
        Defense," *Time*, February 18, 2003, http://content.time.com/time/magazine
        /article/0,9171,423690,00.html; Ron Harris, "Operation Field Chicken Dies,"
        Star News Online, March 2, 2003, https://www.starnewsonline.com/article
        /NC/20030302/News/605075551/WM.

THE ORIGINAL UNDERWATER AGENTS

29   **Doomsday Glacier AUV:** "Scientists Drill for First Time on Remote Antarc-
        tic Glacier," International Thwaites Glacier Collaboration, January 28, 2020,
        https://thwaitesglacier.org/index.php/news/scientists-drill-first-time-remote
        -antarctic-glacier; *Times* Travel Editor, "Scientists Discover Warm Water
        beneath 'Doomsday Glacier' in Antarctica," *Times of India: TimesTravel*, Febru-
        ary 4, 2020, https://timesofindia.indiatimes.com/travel/destinations/scientists
        -discover-warm-water-beneath-doomsday-glacier-in-antarctica/as73901574
        .cms.

30   **Icefin isn't the only robot helping:** Emily Jarvie, "University of Tasma-
        nia Robot Used to Explore under Antarctic Ice," *Examiner*, June 18, 2020,
        https://www.examiner.com.au/story/6797641/antarctic-researchers-explore
        -under-the-ice/.

30   **global market for autonomous underwater vehicles:** "Global Autono-
        mous Underwater Vehicle Market by Vehicle Type, by Technology, by End
        User and by Geography, Forecast and Opportunities, 2025," ReportLinker,
        June 2020, https://www.reportlinker.com/p05916986/Global-Autonomous
        -Underwater-Vehicle-Market-By-Vehicle-Type-By-Technology-By-End
        -User-By-Geography-Forecast-Opportunities.html.

30   **Different uses of AUVs:** University of Exeter, "Robot Cameras Reveal Secret
        Lives of Basking Sharks," *Phys.org*, August 6, 2019, https://phys.org/news
        /2019-08-robot-cameras-reveal-secret-basking.html; "REMUS AUV," Univer-
        sity of Hawai'i at Manoa, accessed August 23, 2020, https://www.soest.hawaii
        .edu/soestwp/tech/watercraft/remus-auv/; Oceanographic Staff, "Remote
        At-Sea Science Expedition Reveals Exciting New Discoveries in Austra-
        lia's Coral Sea Marine Park," *Oceanographic*, June 26, 2020, https://www
        .oceanographicmagazine.com/news/coral-sea-marine-park/; "Surveying
        Deep-Sea Corals, Sponges, and Fish Habitat off the U.S. West Coast," NOAA

Office of Ocean Exploration and Research, accessed August 23, 2020, https://oceanexplorer.noaa.gov/explorations/19express/background/auv/auv.html; Justin Higginbottom, "Underwater Drones Join Hunt for Trillions in Mineral Riches Trapped on Ocean's Floor," CNBC, June 6, 2020, https://www.cnbc.com/2020/06/06/underwater-drones-helping-companies-in-mining-oceans-floor.html.

30 **AUV that can be armed with a special pistol:** Kelsey D. Atherton, "New Russian Robot Will Shoot Naval Mines with a Gun," *Forbes*, June 24, 2020, https://www.forbes.com/sites/kelseyatherton/2020/06/24/new-russian-robot-will-shoot-naval-mines-with-a-gun/?sh=7a2a89e16210.

30 **AUVs have a history of military use:** D. Richard Blidberg, "The Development of Autonomous Underwater Vehicles (AUV): A Brief Summary," 2001, http://wpressutexas.net/cs378h/images/d/de/ICRA_01paper.pdf.

30 **AUVs were used to find and clear mines:** Colin Babb, "NOAA Ocean Explorer: AUVfest 2008; Underwater Mines," National Oceanographic and Atmospheric Administration, https://oceanexplorer.noaa.gov/explorations/08auvfest/background/mines/mines.html.

31 **"self-propelled marine vehicle[s]":** D. C. Morrison, "Marine Mammals Join the Navy," *Science* 242, no. 4885 (December 16, 1988): 1503–4, https://doi.org/10.1126/science.3201237.

31 **Early 2000s marine mammal sale to Iran:** "Iran Buys Kamikaze Dolphins," BBC News, March 8, 2000, http://news.bbc.co.uk/2/hi/middle_east/670551.stm.

31 **British sea lions:** "U-boat Hunting Sea Lions Feature in BBC WW1 E-book," BBC News, October 30, 2015, https://www.bbc.com/news/uk-england-34666242.

31 **CIA report on Soviet dolphins:** "Scientific and Technical Intelligence Report: Capability of the Soviets to Train Marine Mammals for a Military Operational System," Central Intelligence Agency, October 1976, https://www.cia.gov/library/readingroom/docs/DOC_0000969804.pdf.

31 **dolphin helps retrieve lost torpedo:** Nikolai Litovkin, RBTH, and Dmitry Litovkin, "How Russia Is Training a Special Ops Force of Aquatic Mammals," *Russia Beyond*, May 25, 2017, https://www.rbth.com/defence/2017/05/24/how-russia-is-training-a-special-ops-force-of-aquatic-mammals_769405.

31 **BB pellet and a kernel of corn:** William Gasperini, "Uncle Sam's Dolphins," *Smithsonian Magazine*, September 2003, https://www.smithsonianmag.com/science-nature/uncle-sams-dolphins-89811585/.

32 **the Soviets also trained their dolphins to attach mines onto enemy submarines:** Karin Brulliard, "Russia's Military Is Recruiting Dolphins, and Their Mission Is a Mystery," *Washington Post*, March 11, 2016, https://www.washingtonpost.com/news/animalia/wp/2016/03/11/russias-military-is-recruiting-dolphins-and-their-mission-is-a-mystery/.

32   **with harpoons strapped to their heads:** "Iran Buys Kamikaze Dolphins,"
        BBC News.

32   **Dolphins sold as tourist attractions:** Svati Kirsten Narula, "Ukraine Was
        Never Crazy about Its Killer Dolphins, Anyway," *Atlantic*, March 26, 2014,
        https://www.theatlantic.com/international/archive/2014/03/ukraine-was
        -never-crazy-about-its-killer-dolphins-anyway/359647/.

32   **"I am prepared to go to Allah":** "Iran Buys Kamikaze Dolphins," BBC News.

32   **with the goal of phasing in robots by 2017:** Associated Press in San Diego,
        "US Navy's Mine-Hunting Dolphins Will Be Replaced by Robots in 2017,"
        Guardian, December 3, 2012, https://www.theguardian.com/world/2012/dec
        /02/us-military-replace-dolphins-robots.

32   **despite over $90 million in investments:** Matt Potter, "Navy's 'Killer' Dol-
        phins Replaced by Robots? Not So Fast," *San Diego Reader*, November 17,
        2016, https://www.sandiegoreader.com/news/2016/nov/17/ticker-navys
        -killer-dolphins-replaced-robots.

33   **Dolphin hunger strike:** Sam Wolfson, "Ukraine Says Military Dolphins
        Captured by Russia Went on Hunger Strike," *Guardian*, May 17, 2018,
        https://www.theguardian.com/environment/2018/may/16/ukraine-claims
        -dolphin-army-captured-by-russia-went-on-hunger-strike.

33   **2018 *Russia Today* article:** "Военная Служба Ластоногих: В Мурманске
        Тюленей Готовят к Выполнению Боевых Задач," *Russia Today*, February
        16, 2018, https://russian.rt.com/russia/video/482580-voennaya-sluzhba
        -tyuleni-murmansk.

33   **Russian spy whale rumors:** Sebastien Roblin, "The Real 'Submarines':
        Check Out Russia's Combat Dolphins, Spy Whales and Killer Seals," *National
        Interest*, May 4, 2019, https://nationalinterest.org/blog/buzz/real-submarines
        -check-out-russia%E2%80%99s-combat-dolphins-spy-whales-and-killer
        -seals-55667; Hannah Ellis-Petersen, "Whale with Harness Could Be Russian
        Weapon, Say Norwegian Experts," *Guardian*, April 29, 2019, https://www
        .theguardian.com/environment/2019/apr/29/whale-with-harness-could-be
        -russian-weapon-say-norwegian-experts.

33   **"do you really think we'd attach a mobile phone number?":** "Norway Finds
        'Russian Spy Whale' off Coast," BBC News, April 29, 2019, https://www.bbc
        .com/news/world-europe-48090616.

34   **located mines in the Persian Gulf during the Gulf Wars and during the US
        invasion of Iraq:** Alan Boyle, "Dolphins Go to Front Lines in Iraq War," NBC
        News, March 25, 2003, http://www.nbcnews.com/id/3078682/ns/technology
        _and_science-science/t/dolphins-go-front-lines-iraq-war/; Gasperini, "Uncle
        Sam's Dolphins."

AGENTS OF CHANGE

34   **raccoon chimney sweeps:** "Smart Raccoons: Are Trained to Clean Chim-
        neys and Owner Is Getting Rich," *Washington Post*, October 13, 1906.

34   **Silkworms in China, 3000 BCE:** Hui Xiang et al., "The Evolutionary Road from Wild Moth to Domestic Silkworm," *Nature Ecology & Evolution* 2, no. 8 (2018): 1268–79.

34   **Medicinal dogs in ancient Egypt:** Robert S. Kennedy, "Animal-Assisted Therapy: Right for Your Patients?," accessed August 30, 2020, http://neurosciencecme.com/activity/content/664.asp; Stanley Coren, "Foreword," in *Handbook on Animal-Assisted Therapy: Theoretical Foundations and Guidelines for Practice*, ed. Aubrey H. Fine (London: Academic Press, 2010).

34   **we still draw blood from patients using leeches:** Emma Hiolski, "WATCH: See How Leeches Can Be a Surgeon's Sidekick," *Goats and Soda* (blog), NPR, April 3, 2018, https://www.npr.org/sections/goatsandsoda/2018/04/03/598829579/watch-see-how-leeches-can-be-a-surgeons-sidekick.

34   **Domestication of livestock introduces new concepts of ownership, power, and wealth:** Fagan, *Intimate Bond*; Jared Diamond, *Guns, Germs & Steel: A Short History of Everybody for the Last 13,000 Years* (New York: Random House, 2013).

35   **land and structures designed and shaped to support our agricultural pursuits:** Fagan, *Intimate Bond*.

35   **donkeys have been one of the biggest (and underrated) catalysts for change:** Fagan, *Intimate Bond*.

35   **Horses in London; optimize for hay production:** Hannah Velten, *Beastly London: A History of Animals in the City* (London: Reaktion Books, 2013).

2. INTEGRATING THE NEW BREED

37   **Quote:** Douglas Adams, *The Hitchhiker's Guide to the Galaxy* (Swindon: Book Club Associates, 1983).

DIFFERENT KINDS OF INTELLIGENCE

39   **"The Coming Technological Singularity":** Vernor Vinge, "The Coming Technological Singularity: How to Survive in the Post-Human Era," in *Science Fiction Criticism: An Anthology of Essential Writings*, ed. Rob Latham (London: Bloomsbury Academic, 2017), 352–63.

39   **robot vacuum cleaners encountering some dog poop:** Olivia Solon, "Roomba Creator Responds to Reports of 'Poopocalypse': 'We See This a Lot'," *Guardian*, August 15, 2016, https://www.theguardian.com/technology/2016/aug/15/roomba-robot-vacuum-poopocalypse-facebook-post.

39   **Moore's law:** David Rotman, "We're Not Prepared for the End of Moore's Law," *MIT Technology Review*, February 24, 2020, https://www.technologyreview.com/2020/02/24/905789/were-not-prepared-for-the-end-of-moores-law/.

40   **sort farm cucumbers by size, shape, and color:** Kaz Sato, "How a Japanese Cucumber Farmer Is Using Deep Learning and TensorFlow," *Google Cloud*

Blog, August 31, 2016, https://cloud.google.com/blog/products/gcp/how-a
-japanese-cucumber-farmer-is-using-deep-learning-and-tensorflow/.

40   **the parts of the photos that contained human fingers:** Wieland Brendel and
Matthias Bethge, "Approximating CNNs with Bag-of-Local-Features Models
Works Surprisingly Well on ImageNet," arXiv preprint (2019), https://arxiv
.org/abs/1904.00760; Janelle Shane, "When Data Is Messy," *AI Weirdness*, July
3, 2020, https://aiweirdness.com/post/622648824384602112/when-data-is
-messy.

40   **labeling every single image "bolete":** Pedro Tabacof and Eduardo Valle,
"Exploring the Space of Adversarial Images," *2016 International Joint Conference on Neural Networks (IJCNN)*, Vancouver, BC (2016): 426–33, https://doi
.org/10.1109/IJCNN.2016.7727230.

40   **"We just tried to create a machine that could win at *Jeopardy!*":** Gary Smith,
The AI Delusion (Oxford: Oxford University Press, 2018).

41   **narrow and well defined enough to figure out by analyzing a bunch of
data:** Gary Marcus and Ernest Davis, *Rebooting AI: Building Artificial Intelligence We Can Trust* (New York: Pantheon Books, 2019); Gary Marcus and
Ernest Davis, "A.I. Is Harder Than You Think," *New York Times*, May 18,
2018, https://www.nytimes.com/2018/05/18/opinion/artificial-intelligence
-challenges.html; Alice Lloyd George, "Discussing the Limits of Artificial
Intelligence," *TechCrunch*, April 1, 2017, https://techcrunch.com/2017/04/01
/discussing-the-limits-of-artificial-intelligence/.

41   **"if you think it's easy for you to do":** "Ready to Buy a Home Robot?,"
Bloomberg, July 19, 2004, https://www.bloomberg.com/news/articles/2004
-07-18/ready-to-buy-a-home-robot.

41   **"doing broad tasks without even realizing it":** Janelle Shane, *You Look Like
a Thing and I Love You* (London: Hachette UK, 2019).

41   **"this could even turn out to be an impossible problem":** Mark Lee, *How to
Grow a Robot: Developing Human-Friendly, Social AI* (Cambridge, MA: MIT
Press, 2020).

41   **akin to worrying about overpopulation on Mars:** Caleb Garling, "Andrew
Ng: Why 'Deep Learning' Is a Mandate for Humans, Not Just Machines,"
WIRED, May 2015, https://www.wired.com/brandlab/2015/05/andrew-ng
-deep-learning-mandate-humans-not-just-machines/.

42   **"For all we know, human-level intelligence could be a tradeoff":** Maciej
Cegłowski, "Superintelligence: The Idea That Eats Smart People," Keynote:
WebCamp Zagreb Conference 2016, http://2016.webcampzg.org/talks/view
/superintelligence-the-idea-that-eats-smart-people/.

42   **"emu war":** Richard J. Cook and Srđan M. Jovanović, "The Emu Strikes
Back: An Inquiry into Australia's Peculiar Military Action of 1932," *Roma-
nian Journal of Historical Studies* 2, no. 1 (2019).

43   **"Okay, from now on I'll call you 'an ambulance'":** Will Knight, "Tougher
Turing Test Exposes Chatbots' Stupidity," *MIT Technology Review*, July 14,

2016, https://www.technologyreview.com/2016/07/14/7797/tougher-turing -test-exposes-chatbots-stupidity/.

43   **Elephants don't play chess:** Rodney Allen Brooks, "Elephants Don't Play Chess," *Robotics and Autonomous Systems* 6, no. 1–2 (1990): 3–15.

43   **"What Is It Like to Be a Robot?":** Rodney Allen Brooks, "What Is It Like to Be a Robot?," March 18, 2017, https://rodneybrooks.com/what-is-it-like-to -be-a-robot/.

REPLACEMENT VERSUS SUPPLEMENT

43   **Child jockeys:** Syed Mehmood Asghar with Sabir Farhat and Shereen Niaz, "Camel Jockeys of Rahimyar Khan: Findings of a Participatory Research on Life and Situation of Child Camel Jockeys," Save the Children Sweden, 2005, http://lastradainternational.org/lsidocs/351%20Camel-jockeys_of_rahimyar _khan.pdf.

44   **Camel robots:** Jim Lewis, "Robots of Arabia," *WIRED*, November 1, 2005, https://www.wired.com/2005/11/camel/; Ian Sample, "Can Robots Ride Camels?," *Guardian*, April 14, 2005, http://www.theguardian.com/science /2005/apr/14/thisweeksssciencequestions.robots.

44   **"Scavenger" robots in India:** Sonam Joshi, "Sewer-Cleaning Robot Is Saving Lives, but Will It Kill Jobs?," *Times of India*, July 7, 2019, https://timesofindia .indiatimes.com/home/sunday-times/sewer-cleaning-robot-is-saving-lives -but-will-it-kill-jobs/articleshow/70109018.cms.

44   **a number that plummeted to just 2 percent within one century:** David H. Autor, "Skills, Education, and the Rise of Earnings Inequality among the 'Other 99 Percent,'" *Science* 344, no. 6186 (2014): 843–51, https://doi.org/10 .1126/science.1251868.

45   **US Labor Department set up a special committee:** Stanley Levey, "Goldberg Sets Up Office to Cope with Automation; Federal Unit, Which Opens Today, Will Study Manpower and Develop Plans to Retrain Displaced Workers," *New York Times*, April 20, 1961, https://www.nytimes.com/1961/04/20 /archives/goldberg-sets-up-office-to-cope-with-automation-federal-unit -which.html.

45   **2016 study by the VDMA Robotics + Automation Association:** Carolynn Look, "Robots Are Coming, but Not for Your Job," Bloomberg, February 17, 2016, https://www.bloomberg.com/news/articles/2016-02-18/robots-are -coming-but-not-for-your-job.

45   **World Bank's World Development Report:** World Bank, *World Development Report 2019: The Changing Nature of Work*, 2019, https://www .worldbank.org/en/publication/wdr2019.

45   **creates greater labor demand, and generates more wealth:** David H. Autor, "Why Are There Still So Many Jobs? The History and Future of Workplace Automation," *Journal of Economic Perspectives* 29, no. 3 (2015): 3–30.

45   **poised to replace humans in ways we've never seen before:** Martin Ford, *Rise of the Robots: Technology and the Threat of a Jobless Future* (New York: Basic Books, 2015); Erik Brynjolfsson and Andrew McAfee, *The Second Machine Age: Work, Progress, and Prosperity in a Time of Brilliant Technologies* (New York: W. W. Norton, 2014).

45   **When banks introduced ATMs:** James Bessen, "Toil and Technology," *Finance & Development* 52, no. 1 (March 2015), https://www.imf.org/external /pubs/ft/fandd/2015/03/bessen.htm.

46   **most jobs these days require both "labor and capital; brains and brawn":** Autor, "Why Are There Still So Many Jobs?"

46   **"iron horse phenomenon":** David A. Mindell, *Our Robots, Ourselves: Robotics and the Myths of Autonomy* (New York: Viking Adult, 2015).

46   **AI in patent examination:** Gregory Discher, "Artificial Intelligence and the Patent Landscape—Views from the USPTO AI: Intellectual Property Policy Considerations Conference," *Inside Tech Media*, June 14, 2019, https:// www.insidetechmedia.com/2019/06/14/artificial-intelligence-and-the -patent-landscape-views-from-the-uspto-ai-intellectual-property-policy -considerations-conference/.

48   **"No Jobs? Blame the Robots":** "No Jobs? Blame the Robots," *Howard University News Service*, accessed August 24, 2020, http://hunewsservice.com /uncategorized/no-jobs-blame-the-robots/.

48   **the Luddites have been remembered quite unfairly:** Richard Conniff, "What the Luddites Really Fought Against," *Smithsonian Magazine,* March 2011, https://www.smithsonianmag.com/history/what-the-luddites-really -fought-against-264412/.

48   **"the kind of argument that can only be made from a great academic height":** Peter Frase, *Four Futures: Life after Capitalism* (Brooklyn: Verso, 2016).

NEW AGENTS OF CHANGE

49   **Integrate instead of deploy:** Alexandra Mateescu and Madeleine Clare Elish, *AI in Context: The Labor of Integrating New Technologies*, Data & Society, January 30, 2019, https://datasociety.net/library/ai-in-context/.

49   **Driverless trucks?:** Steve Viscelli, "Driverless? Autonomous Trucks and the Future of the American Trucker," University of California, Berkeley, Center for Labor Research and Education, and Working Partnerships USA, September 4, 2018, https://laborcenter.berkeley.edu/driverless/; Karen E. C. Levy, "Digital Surveillance in the Hypermasculine Workplace," *Feminist Media Studies* 16, no. 2 (2016): 361–65, http://dx.doi.org/10.1080/14680777 .2016.1138607; Karen E. C. Levy, "The Contexts of Control: Information, Power, and Truck-Driving Work," *Information Society* 31, no. 2 (2015): 160– 74, https://doi.org/10.1080/01972243.2015.998105; Christophe Haubursin,

"Automation Is Coming for Truckers. But First, They're Being Watched," *Vox*, November 20, 2017, https://www.vox.com/videos/2017/11/20/16670266 /trucking-eld-surveillance.

50   *Player Piano*: Kurt Vonnegut, *Player Piano* (New York: Dial Press, 1952).

51   **household technologies created to reduce women's housework actu- ally increased their unpaid labor:** Ruth Schwartz Cowan, *More Work for Mother: The Ironies of Household Technology from the Open Hearth to the Microwave* (New York: Basic Books, 1983).

51   **"fauxtomation":** Astra Taylor, "The Automation Charade," *Logic*, August 1, 2018, https://logicmag.io/failure/the-automation-charade/; see also Mate- escu and Elish, *AI in Context*.

53   **committee recommended two years of free community college . . . and a guaranteed minimum income:** Harold R. Bowen, *Report of the National Commission on Technology, Automation, and Economic Progress: Volume I* (Washington, DC: US Government Printing Office, 1966).

53   **The Fordist production mind-set that only recognizes value within a narrowly defined concept of labor has not changed:** Taylor, "Automation Charade"; Neferti Tadiar, "If Not Mere Metaphor . . . Sexual Economies Reconsidered," *Scholar & Feminist Online* 7, no. 3 (Summer 2009), http:// sfonline.barnard.edu/sexecon/tadiar_01.htm.

54   **CEOs now make 320 times more than the average worker:** Lawrence Mishel and Jori Kandra, "CEO Compensation Surged 14% in 2019 to $21.3 Million," *Economic Policy Institute*, August 18, 2020, https://www.epi.org /publication/ceo-compensation-surged-14-in-2019-to-21-3-million-ceos -now-earn-320-times-as-much-as-a-typical-worker/.

DESIGN

54   **"This Hot Robot Says She Wants to Destroy Humans":** "This Hot Robot Says She Wants to Destroy Humans," CNBC, March 16, 2016, https://www .cnbc.com/video/2016/03/16/this-hot-robot-says-she-wants-to-destroy -humans.html.

55   **"A Fine Bosom, a Sweet Face and a Computer for a Brain":** Deirdre Falvey, "Meet Sophia: A Fine Bosom, a Sweet Face and a Computer for a Brain," *Irish Times*, January 8, 2020, https://www.irishtimes.com/life-and-style/people/meet -sophia-a-fine-bosom-a-sweet-face-and-a-computer-for-a-brain-1.4134117.

55   **humanoid as the Holy Grail of robotics:** Peter Menzel and Faith D'Aluisio, *Robo Sapiens: Evolution of a New Species* (Cambridge, MA: MIT Press, 2000).

55   **need to look like us because we relate best to other humans:** Yoseph Bar- Cohen and David Hanson, *The Coming Robot Revolution: Expectations and Fears about Emerging Intelligent, Humanlike Machines* (New York: Springer Science & Business Media, 2009).

55   **fascinated with re-creating ourselves:** Elizabeth Broadbent, "Interactions

with Robots: The Truths We Reveal about Ourselves," *Annual Review of Psychology* 68 (2017): 627–52.

56   **"whatever shape gets the job done":** Robin R. Murphy, *Introduction to AI Robotics* (Cambridge, MA: MIT Press, 2019).

56   **they don't need to be human legs:** Lee, *How to Grow a Robot*.

56   **adhesives inspired by sticky gecko feet:** Jia You, "Gecko-Inspired Adhesives Allow People to Climb Walls," *Science*, November 18, 2014, https://www.sciencemag.org/news/2014/11/gecko-inspired-adhesives-allow-people-climb-walls.

56   **mimic self-cooling termite mounds:** "See How Termites Inspired a Building That Can Cool Itself," *National Geographic*, accessed August 31, 2020, https://video.nationalgeographic.com/video/magazine/decoder/00000163-4f96-de63-afe7-7fdf708d0000.

56   **inspired by the wide-angle way eagles keep view of their surroundings:** P. W. Singer, *Wired for War: The Robotics Revolution and Conflict in the 21st Century* (New York: Penguin, 2009).

56   **climb the walls instead of taking the stairs:** David Galbraith, "15 Wall Climbing Robots," Oobject, accessed August 25, 2020, https://www.oobject.com/category/15-wall-climbing-robots/.

57   **Laurel Riek:** Laurel D. Riek, "Healthcare Robotics," *Communications of the ACM* 60, no. 11 (November 2017): 68–78.

URBAN DESIGN AND PUBLIC SPACES

58   **field layouts and broadband internet:** Mateescu and Elish, *AI in Context*.

58   **robots raise urban design questions:** Kristen Thomasen, "Robots, Regulation, and the Changing Nature of Public Spaces," *Ottawa Law Review* 51, no. 2 (2020).

58   **obstructing the path of people who use wheelchairs:** Emily Ackerman, "My Fight with a Sidewalk Robot," Bloomberg, November 19, 2019, https://www.bloomberg.com/news/articles/2019-11-19/why-tech-needs-more-designers-with-disabilities.

58   **using a 400-pound security robot to chase homeless people away:** "San Francisco SPCA Deployed This Security Robot to Chase Off Homeless People," CBC Radio, December 14, 2017, https://www.cbc.ca/radio/asithappens/as-it-happens-thursday-edition-1.4448373/san-francisco-spca-deployed-this-security-robot-to-chase-off-homeless-people-1.4448376.

58   **Architectural choices for shared spaces are . . . made by people with power:** Irene Cheng, Charles L. Davis II, and Mabel O. Wilson, eds., *Race and Modern Architecture: A Critical History from the Enlightenment to the Present* (Pittsburgh: University of Pittsburgh Press, 2020).

3. TRESPASSERS: ASSIGNING RESPONSIBILITY FOR AUTONOMOUS DECISIONS

60   **Quote:** Norbert Wiener, *The Human Use of Human Beings: Cybernetics and Society* (Boston: Houghton Mifflin, 1950).

60   **Quote:** Anonymous case, 12 Henry VII. Keilway 3b.

61   **the individuals on trial in this case were a sow and her piglets:** Edward Payson Evans, *The Criminal Prosecution and Capital Punishment of Animals: The Lost History of Europe's Animal Trials* (London: William Heinemann, 1906); Robert Chambers, *Book of Days*, vol. 1 (London: W. & R. Chambers, 1879); Jacques Berriat Saint-Prix, *Rapport et recherches sur les procès et jugemens relatifs aux animaux* (Paris: Imprimerie de Selligue, 1829).

WARNING: ERROR

61   **Wanda Holbrook case:** Tresa Baldas, "Lawsuit: Defective Robot Killed Factory Worker; Human Error to Blame," *Detroit Free Press*, March 14, 2017, accessed August 25, 2020, https://www.freep.com/story/news/local/michigan/2017/03/14/lawsuit-defective-robot-killed-factory-worker-human-error-blame/99173888/.

62   **robotic surgery has gone awry:** Alessia Ferrarese et al., "Malfunctions of Robotic System in Surgery: Role and Responsibility of Surgeon in Legal Point of View," *Open Medicine* 11, no. 1 (2016): 286–91; see also Balding v. Tarter, 3 N.E.3d 794 (Ill. 2014); Reece v. Intuitive Surgical, Inc., 63 F. Supp. 1337 (N.D. Ala. 2014).

62   **robot knocked over a sixteen-month-old:** James Vincent, "Mall Security Bot Knocks Down Toddler, Breaks Asimov's First Law of Robotics," *The Verge*, July 13, 2016, https://www.theverge.com/2016/7/13/12170640/mall-security-robot-k5-knocks-down-toddler.

62   **search algorithms that reinforce racial or gender biases:** Safiya Umoja Noble, *Algorithms of Oppression: How Search Engines Reinforce Racism* (New York: New York University Press, 2018); see also Leon Yin and Aaron Sankin, "Google Ad Portal Equated 'Black Girls' with Porn," *The Markup*, July 23, 2020, https://themarkup.org/google-the-giant/2020/07/23/google-advertising-keywords-black-girls.

62   **response of virtual assistants to a mental or physical health crisis:** Adam S. Miner et al., "Smartphone-Based Conversational Agents and Responses to Questions about Mental Health, Interpersonal Violence, and Physical Health," *JAMA Internal Medicine* 176, no. 5 (2016): 619–25.

62   **artificial intelligence risk scoring in courtrooms:** "COMPAS Recidivism Risk Score Data and Analysis," *ProPublica Data Store*, November 2020, https://www.propublica.org/datastore/dataset/compas-recidivism-risk

-score-data-and-analysis; Chelsea Barabas, Karthik Dinakar, and Colin Doyle, "The Problems with Risk Assessment Tools," *New York Times*, July 17, 2019, https://www.nytimes.com/2019/07/17/opinion/pretrial-ai.html.

62   **for a deep dive on a myriad of harms in artificial intelligence and its integration:** Noble, *Algorithms of Oppression*; Joy Buolamwini and Timnit Gebru, "Gender Shades: Intersectional Accuracy Disparities in Commercial Gender Classification," *Conference on Fairness, Accountability, and Transparency* (2018); Ruha Benjamin, *Race After Technology: Abolitionist Tools for the New Jim Code* (Cambridge, UK: Polity, 2019); Latanya Sweeney, "Discrimination in Online Ad Delivery," Data Privacy Lab, accessed August 25, 2020, https://dataprivacylab.org/projects/onlineads/index.html; Sasha Costanza-Chock, "Design Justice: Towards an Intersectional Feminist Framework for Design Theory and Practice," in *Proceedings of the Design Research Society* (2018); Cathy O'Neil, *Weapons of Math Destruction: How Big Data Increases Inequality and Threatens Democracy* (New York: Broadway Books, 2016); see also Aura Bogado, "Native Americans Say Facebook Is Accusing Them of Using Fake Names," February 9, 2015, https://www.colorlines.com/articles/native-americans-say-facebook-accusing-them-using-fake-names; Jessica Guynn, "Google Photos Labeled Black People 'Gorillas,'" *USA Today*, July 1, 2015, https://www.usatoday.com/story/tech/2015/07/01/google-apologizes-after-photos-identify-black-people-as-gorillas/29567465/.

63   **intercontinental ballistic missile strike:** Geoffrey Forden, "False Alarms in the Nuclear Age," *PBS Nova*, November 6, 2001, https://www.pbs.org/wgbh/nova/article/nuclear-false-alarms/.

63   **Some describe our current situation as a "historic moment":** Trevor N. White and Seth D. Baum, "Liability for Present and Future Robotics Technology," in *Robot Ethics 2.0: From Autonomous Cars to Artificial Intelligence*, ed. Patrick Lin, Keith Abney, and Ryan Jenkins (Oxford: Oxford University Press, 2017).

63   **about 1.35 million people die in road wrecks:** "Road Safety in the Western Pacific," World Health Organization, accessed August 25, 2020, https://www.who.int/westernpacific/health-topics/road-safety.

63   **whose errors cause more than 90 percent of car accidents:** Bryant Walker Smith, "Human Error as a Cause of Vehicle Crashes," Center for Internet and Society, December 18, 2013, http://cyberlaw.stanford.edu/blog/2013/12/human-error-cause-vehicle-crashes; see also "Automated Vehicles for Safety," National Highway Traffic Safety Administration, United States Department of Transportation, accessed August 30, 2020, https://www.nhtsa.gov/technology-innovation/automated-vehicles-safety.

64   **only 43 percent of Americans are even interested in using self-driving cars:** "Allianz Global Assistance Survey Finds Fewer Americans Interested in Autonomous Vehicles," Green Car Congress, September 19, 2018, https://www.greencarcongress.com/2018/09/20180919-allianz.html.

64   **56 percent wouldn't even want to ride in a driverless car if given the chance:**
        Aaron Smith and Monica Anderson, "Automation in Everyday Life," Pew
        Research Center, October 4, 2017, https://www.pewresearch.org/internet
        /2017/10/04/automation-in-everyday-life/.

64   **throwing rocks at them and slashing their tires:** Simon Romero, "Wielding
        Rocks and Knives, Arizonans Attack Self-Driving Cars," *New York Times*,
        December 31, 2018, https://www.nytimes.com/2018/12/31/us/waymo-self
        -driving-cars-arizona-attacks.html.

64   **Asimov's three laws of robotics:** Isaac Asimov, "Runaround," *Astounding
        Science Fiction* 29.1 (1942): 94–103.

65   **Maja Mataric, "Which tells you something about [their] role in the field":**
        Lee McCauley, "The Frankenstein Complex and Asimov's Three Laws," in
        Proceedings of the Papers from the 2007 AAAI Workshop (Menlo Park, CA:
        AAAI Press, 2007).

65   **translating those ethics into code has proven a seemingly impossible task:**
        Colin Allen and Wendel Wallach, "Moral Machines: Contradiction in Terms,
        or Abdication of Human Responsibility," in *Robot Ethics: The Ethical and
        Social Implications of Robotics*, ed. Patrick Lin, Keith Abney, and George A.
        Bekey (Cambridge, MA: MIT Press, 2012), 55–68.

66   **"is not and cannot be the attempt to re-create full moral agency":** Allen
        and Wallach, "Moral Machines."

THE CASE OF THE GORING OX

67   **cattle are among the most important animals we cultivate:** Brian Fagan,
        The Intimate Bond: How Animals Shaped Human History (New York:
        Bloomsbury, 2015).

68   **some of the earliest extant "laws" known to humankind:** Reuven Yaron,
        The Laws of Eshnunna (Leiden, Netherlands: Brill, 1988).

68   **first known rules created by Mesopotamian societies that lay out clear
        consequences for when an animal causes harm:** J. J. Finkelstein, "The Ox
        That Gored," *Transactions of the American Philosophical Society* 71, no. 2
        (1981): 1–89; Steven M. Wise, "The Legal Thinghood of Nonhuman Ani-
        mals," *Boston College Environmental Affairs Law Review* 23, no. 3 (1996): 471.

68   **noxal surrender:** Stefan Jurasinski, "Noxal Surrender, the 'Deodand,' and
        the Laws of King Alfred," *Studies in Philology* 111, no. 2 (2014): 195–224.

68   **deodand:** Anna Pervukhin, "Deodands: A Study in the Creation of Common
        Law Rules," *American Journal of Legal History* 47, no. 3 (July 2005): 237–56.

CATEGORIES AND FENCES

70   **"for I am the trespasser with my beasts":** Anonymous case, 12 Henry VII.
        Keilway 3b.

70   **Pigs wandered the streets:** Fagan, *Intimate Bond.*

70   **even if their fences were perfectly built:** Jacob Henry Beuscher, *Law and the Farmer*, 3rd ed. (New York: Springer, 1960); Sonia Waisman, Bruce A. Wagman, and Pamela D. Frasch, *Animal Law: Cases and Materials*, 2nd ed. (Durham, NC: Carolina Academic Press, 2002).

70   **landowner's responsibility to build a fence:** Beuscher, *Law and the Farmer.*

70   **in 1926 in Wisconsin, a court decided:** Fox v. Koenig, 190 Wis. 528, 209 N. W. 708 (1926).

70   **"moved by some wildness contrary to the nature of its kind":** Wise, "Legal Thinghood," 471; Dig. 9.1.7 (Ulpian, Edict, book 18).

71   **held to task for damage caused by their trespassing canines:** Beuscher, *Law and the Farmer.*

71   **the Bedouin have tamed gazelles:** Richard W. Bulliet, *Hunters, Herders, and Hamburgers: The Past and Future of Human-Animal Relationships* (New York: Columbia University Press, 2005).

71   **the laws on bees:** Bruce W. Frier, "Bees and Lawyers," *Classical Journal* 78, no. 2 (1982): 105–14; Harry R. Trusler, "The Law of Bees," *North Carolina Law Review* 5 (1926): 46; E. J. Cohn, "Bees and the Law," *Law Quarterly Review* 218 (1939): 289–94; German civil code (§§960-61); Lucas v. Pettit, 12 Ont. L. Rep. 448; Anthony Rago, "Bees and the Law: What You Need to Know," Hobby Farms, November 16, 2015, https://www.hobbyfarms.com/bees-and-the-law -what-you-need-to-know-2/.

72   **it's not an inherently dangerous technology like an autonomous car:** Peter M. Asaro, "What Should We Want from a Robot Ethic?," *International Review of Information Ethics* 6 (2006): 9–16.

72   **federal robotics commission:** Ryan Calo, *The Case for a Federal Robotics Commission*, Brookings, September 15, 2014, https://www.brookings.edu /research/the-case-for-a-federal-robotics-commission/.

DOG LICENSES AND SHEEP FUNDS

72   **Wiener Wiesn police dog incident:** "Betrunkener riss Polizeihund in Wien Maulkorb herunter und wurde gebissen," *Der Standard*, accessed August 26, 2020, https://www.derstandard.at/story/2000088818327/betrunkener-riss -polizeihund-in-wien-maulkorb-herunter-und-wurde-gebissen.

73   **Vienna dangerous dog rules:** "Listenhunde ('Kampfhunde')—Übersicht und Vorgaben," accessed August 26, 2020, https://www.wien.gv.at/gesellschaft /tiere/haustiere/hunde/listenhunde.html.

74   **We could implement rules and design standards for robots:** See also Enrique Schaerer, Richard Kelley, and Monica Nicolescu, "Robots as Animals: A Framework for Liability and Responsibility in Human-Robot Interactions," in *RO-MAN 2009—The 18th IEEE International Symposium on Robot and Human Interactive Communication*, Toyama (2009): 72–77.

74   **English tax on dogs in 1796:** Fagan, *Intimate Bond.*

74   **"sheep funds":** Beuscher, *Law and the Farmer.*

75   **State of Ohio coyote law:** Ohio code 955.51 Claims for value of animals injured or killed by coyote. Effective Dates: September 26, 2003; April 15, 2005; see also Ohio code 955.52–53.

THE ANIMAL TRIALS

76   **the ox that fatally gored a person was stoned to death:** Wise, "Legal Thinghood," 471; Finkelstein, "The Ox That Gored," 1–89; see also chapters 21–23:19 of the book of Exodus.

76   **religious laws of the Vendidad:** Walter Woodburn Hyde, "The Prosecution and Punishment of Animals and Lifeless Things in the Middle Ages and Modern Times," *University of Pennsylvania Law Review and American Law Register* 64, no. 7 (1916): 696–730.

76   **animals stand trial before twenty-three judges:** Finkelstein, "The Ox That Gored"; Tractate Sanhedrin 1.4, in the Mishnah, trans. Paul Selver (1964): 383.

76   **the Greeks also put animals on trial for murder:** Marilyn A. Katz, "Ox-Slaughter and Goring Oxen: Homicide, Animal Sacrifice, and Judicial Process," *Yale Journal of Law & the Humanities* 4, no. 2 (1992): 249.

76   **Basel rooster trial:** Johann Georg Groß, *Kurze Basler Chronik: von XIV hundert Jahren bis auf das Jahr 1624.*

77   **to a lesser extent in Spain, Denmark, Turkey, and the Scandinavian countries, as well as in other parts of the world:** Thomas G. Kelch, "A Short History of (Mostly) Western Animal Law: Part I," *Animal Law* 19 (2012): 23; Wise, "Legal Thinghood of Nonhuman Animals."

77   **the most frequent criminal was the pig:** Evans, *Criminal Prosecution*; Hyde, "Prosecution and Punishment."

77   **generally a secular proceeding:** Karl von Amira, "Thierstrafen und thierprocesse," *Mitteilungen des Instituts für Oesterreichische Geschichtsforschung* 12, no. 1 (1891): 545–601.

77   **the church started performing another type of animal trial:** Kelch, "A Short History of (Mostly) Western Animal Law."

77   **When weevils infested the vineyards of St. Julien:** Evans, *Criminal Prosecution and Capital Punishment of Animals.*

77   **how meticulously they followed the same legal process as for humans:** Kelch, "A Short History of (Mostly) Western Animal Law."

78   **read out the summons in public areas where the creatures were most likely to hear it:** Hyde, "Prosecution and Punishment."

78   **Bartholomy de Chassenée:** William Ewald, "Comparative Jurisprudence (I): What Was It Like to Try a Rat?," *University of Pennsylvania Law Review* 143, no. 6 (1995): 1889–2149; Evans, *Criminal Prosecution and Capital Punishment of Animals*; Sadakat Kadri, *The Trial: Four Thousand Years of Courtroom Drama* (New York: Random House, 2007); Hampton L. Carson, "The Trial of Animals and Insects, a Little Known Chapter of Mediæval

Jurisprudence," in *Proceedings of the American Philosophical Society* (1917): 410–15.

79   **Mary the circus elephant:** Thomas G. Burton, "The Hanging of Mary: A Circus Elephant," *Tennessee Folklore Society Bulletin* 37 (1971): 1–8; Hilda Padgett, "The Hanging of Mary the Elephant," US Gen Web Project, Erwin, Tennessee (1996), accessed August 27, 2020, https://sites.rootsweb.com /~tnunicoi/mary.htm.

79   **the Commonwealth of Kentucky sentenced a German shepherd to death:** Geoffrey P. Goodwin and Adam Benforado, "Judging the Goring Ox: Retribution Directed Toward Animals," *Cognitive Science* 39, no. 3 (2015): 619–46.

THE NEW BLAME GAME

80   **opting instead for images of metal humanoid robots to illustrate:** See, for example, Ephrat Livni, "A Rogue Robot Is Blamed for a Human Colleague's Gruesome Death," *Quartz*, March 13, 2017, https://qz.com/931304/a-robot -is-blamed-in-death-of-a-maintenance-technician-at-ventra-ionia-main-in -michigan/.

80   **report by the European Parliament's Committee on Legal Affairs:** *Report with Recommendations to the Commission on Civil Law Rules on Robotics*, Committee on Legal Affairs, January 27, 2017, https://www.europarl.europa .eu/doceo/document/A-8-2017-0005_EN.html; Rachel Withers, "The EU Is Trying to Decide Whether to Grant Robots Personhood," *Slate*, April 17, 2018, https://slate.com/technology/2018/04/the-eu-is-trying-to-decide -whether-to-grant-robots-personhood.html.

80   **Many an author, legal scholar, or philosopher has argued:** John Frank Weaver, *Robots Are People Too: How Siri, Google Car, and Artificial Intelligence Will Force Us to Change Our Laws* (Santa Barbara, CA: Praeger, 2013); David C. Vladeck, "Machines without Principals: Liability Rules and Artificial Intelligence," *Washington Law Review* 89 (2014): 117; John P. Sullins, "Ethics and Artificial Life: From Modeling to Moral Agents," *Ethics and Information Technology* 7, no. 3 (2005): 139.

81   **In 2018, the US Chamber Institute for Legal Reform published a report:** *Torts of the Future II: Addressing the Liability and Regulatory Implications of Emerging Technologies*, US Chamber Institute for Legal Reform, April 2018, https://instituteforlegalreform.com/research/torts-of-the-future -ii-addressing-the-liability-and-regulatory-implications-of-emerging -technologies/.

81   **A small slice of overshadowed academic work:** See, for example, Schaerer, Kelley, and Nicolescu, "Robots as Animals," 72–77; Richard Kelley et al., "Liability in Robotics: An International Perspective on Robots as Animals," *Advanced Robotics* 24, no. 13 (2010): 1861–71.

82   **we'll often look to the companies who make and sell them:** Jeffrey Gurney, "Imputing Driverhood: Applying a Reasonable Driver Standard to Accidents Caused by Autonomous Vehicles," in *Robot Ethics 2.0: From Autonomous Cars to Artificial Intelligence*, ed. Patrick Lin, Keith Abney, and Ryan Jenkins (Oxford: Oxford University Press, 2016).

82   **where a robotic platform uses open source software with many anonymous contributors:** Ryan Calo, "Open Robotics," *Maryland Law Review* 70, no. 3 (2011): 571.

83   **because the robots themselves have developed the ability to suffer:** Gert-Jan Lokhorst and Jeroen van den Hoven, "Responsibility for Military Robots," in *Robot Ethics: The Ethical and Social Implications of Robotics*, ed. Patrick Lin, Keith Abney, and George A. Bekey (Cambridge, MA: MIT Press, 2012), 145–56.

83   **satisfies people's desire for justice and retribution:** Christina Mulligan, "Revenge Against Robots," *South Carolina Law Review* 69, no. 3 (2018): 579.

83   **Android Fallacy; both they and legal scholar Ryan Calo have cautioned:** Neil M. Richards and William D. Smart, "How Should the Law Think about Robots?," in *Robot Law*, ed. Ryan Calo, A. Michael Froomkin, and Ian R. Kerr (Cheltenham, UK: Edward Elgar, 2016); Ryan Calo, "Robots in American Law," *University of Washington School of Law Research Paper* 2016-04 (2016).

83   **to prevent people from escaping responsibility for their actions:** Asaro, "What Should We Want from a Robot Ethic?"

83   **people try to blame the robot with AI systems in play today:** John Villasenor, *Products Liability Law as a Way to Address AI Harms*, Brookings, October 31, 2019, https://www.brookings.edu/research/products-liability-law-as-a-way-to-address-ai-harms/; see also Sean Coughlan, "A-levels and GCSEs: Boris Johnson Blames 'Mutant Algorithm' for Exam Fiasco," BBC News, August 26, 2020, https://www.bbc.com/news/education-53923279.

84   **When Microsoft released a Twitter bot named Tay:** James Vincent, "Twitter Taught Microsoft's AI Chatbot to Be a Racist Asshole in Less Than a Day," *The Verge*, March 24, 2016, https://www.theverge.com/2016/3/24/11297050/tay-microsoft-chatbot-racist.

HUMAN IN THE LOOP

85   **most recently, the back of a police car:** Meira Gebel, "Tesla Driver Using Autopilot Suspected of DUI after Crashing into Arizona Cop Car," *Digital Trends*, July 14, 2020, https://www.digitaltrends.com/cars/tesla-driver-using-autopilot-suspected-of-dui-after-crashing-into-arizona-state-trooper/.

85   **the media tends to attribute the plane crashes to the human pilots:** Madeleine Clare Elish and Tim Hwang, "Praise the Machine! Punish the Human! The Contradictory History of Accountability in Automated Aviation"

(working paper, Data & Society, May 18, 2015, revised August 12, 2017), https://papers.ssrn.com/sol3/papers.cfm?abstract_id=2720477.

85 **"moral crumple zone":** Madeleine Clare Elish, "Moral Crumple Zones: Cautionary Tales in Human-Robot Interaction," *Engaging Science, Technology, and Society* 5 (2019): 40–60.

II COMPANIONSHIP

4. ROBOTS VERSUS TOASTERS

89 **Quote:** Aristotle, *Politika*, ca. 328 BCE.

89 **Around 2014, Hiroshi Funabashi received an odd request:** Elena Knox and Katsumi Watanabe, "AIBO Robot Mortuary Rites in the Japanese Cultural Context," in *2018 IEEE/RSJ International Conference on Intelligent Robots and Systems (IROS 2018)*: 2020–25; James Burch, "In Japan, a Buddhist Funeral Service for Robot Dogs," *National Geographic*, May 24, 2018, https://www.nationalgeographic.com/travel/destinations/asia/japan/in-japan--a-buddhist-funeral-service-for-robot-dogs/.

90 **selling 150,000 units worldwide:** Takashi Mochizuki and Eric Pfanner, "In Japan, Dog Owners Feel Abandoned as Sony Stops Supporting 'Aibo,'" *Wall Street Journal*, February 11, 2015, https://www.wsj.com/articles/in-japan-dog-owners-feel-abandoned-as-sony-stops-supporting-aibo-1423609536.

90 **Sony announced that they would be discontinuing AIBO:** John Borland, "Sony Puts Aibo to Sleep," *CNET*, January 27, 2006, https://www.cnet.com/news/sony-puts-aibo-to-sleep/.

90 **AIBO funerals:** Scott Neuman, "In Japan, Old Robot Dogs Get a Buddhist Send-Off," NPR, May 1, 2018, https://www.npr.org/sections/thetwo-way/2018/05/01/607295346/in-japan-old-robot-dogs-get-a-buddhist-send-off; Miwa Suzuki, "In Japan, Aibo Robots Get Their Own Funeral," *Japan Times*, May 1, 2018, https://web.archive.org/web/20180502031557if_/https:/www.japantimes.co.jp/news/2018/05/01/national/japan-aibo-robots-get-funeral/#.Wukti5P7TUI; Knox and Watanabe, "AIBO Robot Mortuary Rites in the Japanese Cultural Context"; Burch, "In Japan, a Buddhist Funeral Service for Robot Dogs."

90 **went on to explore a new venture called Robot Therapy:** Knox and Watanabe, "AIBO Robot Mortuary Rites in the Japanese Cultural Context."

91 **holding funeral ceremonies for their possessions:** Angelika Kretschmer, "Mortuary Rites for Inanimate Objects: The Case of Hari Kuyō," *Japanese Journal of Religious Studies* 27, no. 3/4 (2000): 379–404.

91 **"I thought my head was going to explode":** Leander Kahney, "Puppy Love for a Robot," *WIRED*, February 22, 2001, https://www.wired.com/2001/02/puppy-love-for-a-robot/.

ANTHROPOMORPHISM (PROJECTING OURSELVES)

92   **Rosie is one of the top names:** Daniel Aleksandersen, "Top 15 Roomba Names", *Ctrl Blog*, March 30, 2018, https://www.ctrl.blog/entry/popular -roomba-names.html.

92   **we've been drawn to the idea of autonomous "robots" for millennia:** Adrienne Mayor, *Gods and Robots: Myths, Machines, and Ancient Dreams of Technology* (Princeton, NJ: Princeton University Press, 2020).

92   **"indistinguishable from magic":** Arthur C. Clarke, *Profiles of the Future* (London: Hachette UK, 2013).

93   **even adults constantly assign human traits to animals and objects:** Carolyn L. Burke and Joby G. Copenhaver, "Animals as People in Children's Literature," *Language Arts* 81, no. 3 (2004): 205–13.

93   **Gary Dahl became a multimillionaire by selling pet rocks:** James Stuart Olson, ed., *Historical Dictionary of the 1970s* (Westport, CT: Greenwood Press, 1999).

93   **in the Middle to Upper Paleolithic transition, some 40,000 years ago:** Steven Mithen, "The Prehistory of the Mind," *Cambridge Archaeological Journal* 7 (1997): 269; James A. Serpell, "Anthropomorphism and Anthropomorphic Selection—Beyond the 'Cute Response,'" *Society & Animals* 11, no. 1 (2003): 83–100.

93   **needing to form social alliances to survive:** Nicholas Epley, Adam Waytz, and John T. Cacioppo, "On Seeing Human: A Three-Factor Theory of Anthropomorphism," *Psychological Review* 114, no. 4 (2007): 864; Nicholas Epley et al., "When We Need a Human: Motivational Determinants of Anthropomorphism," *Social Cognition* 26, no. 2 (2008): 143–55.

93   **scientists just assuming there must be *some* sort of reason for it:** Alexandra C. Horowitz and Marc Bekoff, "Naturalizing Anthropomorphism: Behavioral Prompts to Our Humanizing of Animals," *Anthrozoös* 20, no. 1 (2007): 23–35.

93   **robust over time and already present in young children:** Adam Waytz, John T. Cacioppo, and Nicholas Epley, "Who Sees Human? The Stability and Importance of Individual Differences in Anthropomorphism," *Perspectives on Psychological Science* 5, no. 3 (2010): 219–32; Ceylan Özdem et al., "Believing Androids—fMRI Activation in the Right Temporo-Parietal Junction Is Modulated by Ascribing Intentions to Non-Human Agents," *Social Neuroscience* 12, no. 5 (2017): 582–93.

94   **even just a black-and-white drawing of facial features:** Robert L. Fantz, "The Origin of Form Perception," *Scientific American* 204, no. 5 (1961): 66–73; Nancy Kanwisher, Josh McDermott, and Marvin M. Chun, "The Fusiform Face Area: A Module in Human Extrastriate Cortex Specialized for Face Perception," *Journal of Neuroscience* 17, no. 11 (1997): 4302–11.

94   **Babies also seem able to create simulations:** Andrew N. Meltzoff and Rechele Brooks, "'Like Me' as a Building Block for Understanding Other Minds:

Bodily Acts, Attention, and Intention," in *Intentions and Intentionality: Foundations of Social Cognition*, ed. Bertram F. Malle, Louis J. Moses, and Dare A. Baldwin (Cambridge, MA: MIT Press, 2001), 171–91.

94   **from childhood, perceive animals and certain objects as social subjects:** Olin Eugene Myers Jr. and Carol D. Saunders, "Animals as Links toward Developing Caring Relationships with the Natural World," in *Children and Nature: Psychological, Sociocultural, and Evolutionary Investigations*, ed. Peter H. Kahn Jr. and Stephen R. Kellert (Cambridge, MA: MIT Press, 2002), 153–78.

94   **helps us understand the world around us:** Epley, Waytz, and Cacioppo, "On Seeing Human," 864; Adam Waytz, Nicholas Epley, and John T. Cacioppo, "Social Cognition Unbound: Insights into Anthropomorphism and Dehumanization," *Current Directions in Psychological Science* 19, no. 1 (2010): 58–62.

94   **objectophilia:** See, for example, Mark D. Griffiths, "Intimate and Inanimate," *Psychology Today*, July 25, 2013, https://www.psychologytoday.com/us/blog /in-excess/201307/intimate-and-inanimate.

95   **Tamagotchis:** See, for example, Neil West, "Finding Companionship in a Digital Age," *Next Generation Magazine* 34 (1997): 57–63.

95   **some players chose to forfeit their victory:** Pedro Peres, "Apex Data Miner Finds a Gun Charm That Shows Wattson Holding a Companion Cube," *Dot Esports*, June 26, 2020, https://dotesports.com/apex-legends/news/apex -data-miner-finds-a-gun-charm-that-shows-wattson-holding-a-companion -cube.

95   **In the late 1990s, some Stanford researchers:** Clifford Nass, Youngme Moon, and Paul Carney, "Are Respondents Polite to Computers? Social Desirability and Direct Responses to Computers," *Journal of Applied Social Psychology* 29, no. 5 (1999): 1093–110.

96   **Nass and Reeves coined the term "the media equation":** Byron Reeves and Clifford Nass, *The Media Equation: How People Treat Computers, Television, and New Media Like Real People and Places* (Cambridge: Cambridge University Press, 1996).

96   **book of the cartoon paper clip engaging in lewd acts:** Leonard Delaney, "Conquered by Clippy: An Erotic Short Story" (self-pub., 2015).

96   **Nass's research helped Microsoft figure out what went wrong:** Clifford Nass and Corina Yen, *The Man Who Lied to His Laptop: What We Can Learn about Ourselves from Our Machines* (New York: Penguin, 2010).

THE POWER OF MOVEMENT

97   **this story is no more than an urban legend:** Martin Loiperdinger, "Lumière's Arrival of the Train: Cinema's Founding Myth," trans. Bernd Elzer, *Moving Image* 4, no. 1 (2004): 89–118.

98   **seminal study from the 1940s:** Fritz Heider and Marianne Simmel, "An

Experimental Study of Apparent Behavior," *American Journal of Psychology* 57, no. 2 (1944): 243–59.

99   **autonomous movement activates our "life-detector":** Francesca Simion et al., "The Processing of Social Stimuli in Early Infancy: From Faces to Biological Motion Perception," *Progress in Brain Research* 189 (2011): 173–93; Nikolaus F. Troje and Cord Westhoff, "The Inversion Effect in Biological Motion Perception: Evidence for a 'Life Detector'?," *Current Biology* 16, no. 8 (2006): 821–24; Mark H. Johnson, "Biological Motion: A Perceptual Life Detector?," *Current Biology* 16, no. 10 (2006): R376–R377.

99   **Researchers Joshua New, Leda Cosmides, and John Tooby showed people:** Joshua New, Leda Cosmides, and John Tooby, "Category-Specific Attention for Animals Reflects Ancestral Priorities, Not Expertise," in *Proceedings of the National Academy of Sciences* 104, no. 42 (2007): 16598–603.

99   **even stronger when it has a body and is in the room with us:** See, for example, Wilma A. Bainbridge et al., "The Benefits of Interactions with Physically Present Robots over Video-Displayed Agents," *International Journal of Social Robotics* 3, no. 1 (2011): 41–52.

99   **tested this projection with a moving stick:** John Harris and Ehud Sharlin, "Exploring the Affect of Abstract Motion in Social Human-Robot Interaction," *2011 RO-MAN*, Atlanta, GA (2011): 441–48.

100   **will follow its gaze, mimic its behavior, and be more willing to take the physical robot's advice:** Younbo Jung and Kwan Min Lee, "Effects of Physical Embodiment on Social Presence of Social Robots," in *Proceedings of PRESENCE* (2004): 80–87; Galit Hofree, Paul Ruvolo, Marian Stewart Bartlett, and Piotr Winkielman, "Bridging the Mechanical and the Human Mind: Spontaneous Mimicry of a Physically Present Android," *PLOS ONE* 9, no. 7 (2014): e99934; Kyveli Kompatsiari, Vadim Tikhanoff, Francesca Ciardo, Giorgio Metta, and Agnieszka Wykowska, "The Importance of Mutual Gaze in Human-Robot Interaction," *International Conference on Social Robotics* (2017): 443–52; Sara Kiesler, Aaron Powers, Susan R. Fussell, and Cristen Torrey, "Anthropomorphic Interactions with a Robot and Robot-like Agent," *Social Cognition* 26, no. 2 (2008): 169–81; Jamy Li, "The Benefit of Being Physically Present: A Survey of Experimental Works Comparing Copresent Robots, Telepresent Robots and Virtual Agents," *International Journal of Human-Computer Studies* 77 (2015): 23–37.

100   **more willing to obey orders from a physical robot:** Bainbridge et al.," Benefits of Interactions with Physically Present Robots."

100   **people cheat less when a robot is with them:** Guy Hoffman, Jodi Forlizzi, Shahar Ayal, Aaron Steinfeld, John Antanitis, Guy Hochman, Eric Hochendoner, and Justin Finkenaur, "Robot Presence and Human Honesty: Experimental Evidence," in *10th ACM/IEEE International Conference on Human-Robot Interaction (HRI) (2015)*: 181–88.

100   **children learn more from working with a robot:** See, for example, Daniel Leyzberg, Samuel Spaulding, Mariya Toneva, and Brian Scassellati, "The

Physical Presence of a Robot Tutor Increases Cognitive Learning Gains," in *Proceedings of the Annual Meeting of the Cognitive Science Society* 3, no. 34 (2012).

100   **recognizing a robot's emotional cues:** Nicole Lazzeri, Daniele Mazzei, Alberto Greco, Annalisa Rotesi, Antonio Lanatà, and Danilo Emilio De Rossi, "Can a Humanoid Face Be Expressive? A Psychophysiological Investigation," *Frontiers in Bioengineering and Biotechnology* 3 (2015): 64.

100   **empathize more with physical robots:** Stela H. Seo, Denise Geiskkovitch, Masayuki Nakane, Corey King, and James E. Young, "Poor Thing! Would You Feel Sorry for a Simulated Robot? A Comparison of Empathy toward a Physical and a Simulated Robot," in *10th ACM/IEEE International Conference on Human-Robot Interaction (HRI)* (2015): 125–32; Sonya S. Kwak, Yunkyung Kim, Eunho Kim, Christine Shin, and Kwangsu Cho, "What Makes People Empathize with an Emotional Robot?: The Impact of Agency and Physical Embodiment on Human Empathy for a Robot," in *IEEE RO-MAN* (2013): 180–85.

100   **when researchers told children to put a robot in a closet:** Peter H. Kahn Jr., Takayuki Kanda, Hiroshi Ishiguro, Nathan G. Freier, Rachel L. Severson, Brian T. Gill, Jolina H. Ruckert, and Solace Shen, "'Robovie, You'll Have to Go into the Closet Now': Children's Social and Moral Relationships with a Humanoid Robot," *Developmental Psychology* 48, no. 2 (2012): 303.

100   **adults will hesitate to switch off or hit a robot:** Christoph Bartneck, Michel Van Der Hoek, Omar Mubin, and Abdullah Al Mahmud, "'Daisy, Daisy, Give Me Your Answer Do!' Switching Off a Robot," in *2nd ACM/IEEE International Conference on Human-Robot Interaction (HRI)* (2007): 217–22; Christoph Bartneck, Marcel Verbunt, Omar Mubin, and Abdullah Al Mahmud, "To Kill a Mockingbird Robot," in *Proceedings of the ACM/IEEE International Conference on Human-Robot Interaction* (2007): 81–87.

100   **People are polite to robots and try to help them:** Matthias Rehm and Anders Krogsager, "Negative Affect in Human Robot Interaction: Impoliteness in Unexpected Encounters with Robots," in *IEEE RO-MAN* (2013): 45–50.

100   **are friendlier if a robot greets them first:** Min Kyung Lee, Sara Kiesler, and Jodi Forlizzi, "Receptionist or Information Kiosk: How Do People Talk with a Robot?," in *Proceedings of the 2010 ACM Conference on Computer Supported Cooperative Work* (2010): 31–40.

100   **People reciprocate when robots help them:** Lara Lammer, Andreas Huber, Astrid Weiss, and Markus Vincze, "Mutual Care: How Older Adults React When They Should Help Their Care Robot," in *AISB2014: Proceedings of the 3rd International Symposium on New Frontiers in Human-Robot Interaction* (London: Routledge, 2014), 1–4; Eduardo Benítez Sandoval, Jürgen Brandstetter, Mohammad Obaid, and Christoph Bartneck, "Reciprocity in Human-Robot Interaction: A Quantitative Approach through the Prisoner's Dilemma and the Ultimatum Game," *International Journal of Social Robotics* 8, no. 2 (2016): 303–17.

100   **when people don't like a robot, they will call it names:** Rehm and Krog-sager, "Negative Affect in Human Robot Interaction."

100   **By 2004, a million of them had been deployed:** "iRobot's Roomba Robotic Floorvac Surpasses 1 Million in Sales," iRobot, October 25, 2004, https://investor.irobot.com/news-releases/news-release-details/irobots-roomba-robotic-floorvac-surpasses-1-million-sales.

101   **"the new pet craze":** Leander Kahney, "The New Pet Craze: Robovacs," *WIRED*, June 16, 2003, https://www.wired.com/2003/06/the-new-pet-craze-robovacs/.

101   **A 2007 study found:** Ja-Young Sung, Lan Guo, Rebecca E. Grinter, and Henrik I. Christensen, "'My Roomba Is Rambo': Intimate Home Appliances," *International Conference on Ubiquitous Computing* (Berlin: Springer, 2007): 145–62.

BIOLOGICALLY INSPIRED DESIGN

102   **Archytas of Tarentum created a flying wooden pigeon:** William Keith Chambers Guthrie, *A History of Greek Philosophy*, vol. 1, *The Earlier Presocratics and the Pythagoreans* (Cambridge: Cambridge University Press, 1962); John W. Humphrey, John Peter Oleson, Andrew Neil Sherwood, and Milo Nikolic, *Greek and Roman Technology: A Sourcebook; Annotated Translations of Greek and Latin Texts and Documents* (London: Psychology Press, 1998).

103   **PigeonBot:** Eric Chang, Laura Y. Matloff, Amanda K. Stowers, and David Lentink, "Soft Biohybrid Morphing Wings with Feathers Underactuated by Wrist and Finger Motion," *Science Robotics* 5, no. 38 (2020).

103   **directional Velcro:** Evan Ackerman, "PigeonBot Uses Real Feathers to Explore How Birds Fly," *IEEE Spectrum*, January 16, 2020, https://spectrum.ieee.org/automaton/robotics/drones/pigeonbot-uses-real-feathers-to-explore-how-birds-fly.

105   **"Jibo isn't an appliance, it's a companion":** Lance Ulanoff, "Jibo Wants to Be the World's First Family Robot," *Mashable*, July 16, 2014, https://mashable.com/2014/07/16/jibo-worlds-first-family-robot/.

105   **$25 million in series A investments:** Aaron Tilley, "Family Robot Jibo Raises $25 Million in Series A Round," *Forbes*, January 21, 2015, https://www.forbes.com/sites/aarontilley/2015/01/21/family-robot-jibo-raises-25-million/.

106   **"When robots make eye contact":** Sherry Turkle, "In Good Company? On the Threshold of Robotic Companions," in *Close Engagements with Artificial Companions: Key Social, Psychological, Ethical and Design Issues*, ed. Yorick Wilks (Amsterdam: John Benjamins, 2010), 3–10.

SOCIAL ROBOT DESIGN: LESS IS MORE

106   **The ancient Greeks used automated mannequins:** Humphrey et al., *Greek and Roman Technology*.

106   **Leonardo da Vinci built a mechanical knight:** Mark Rosheim, *Leonardo's Lost Robots* (Berlin: Springer Science & Business Media, 2006).

106   **Karakuri puppets:** Yasuhiro Yokota, "A Historical Overview of Japanese Clocks and Karakuri," in *International Symposium on History of Machines and Mechanisms* (New York: Springer, 2009): 175–88.

107   **Lions were kept in cages thousands of years BCE:** Vernon N. Kisling Jr., ed., *Zoo and Aquarium History: Ancient Animal Collections to Zoological Gardens* (Boca Raton, FL: CRC Press, 2000).

107   **claim that the ideal social robot looks and behaves just like a human:** Yoseph Bar-Cohen and David Hanson, *The Coming Robot Revolution: Expectations and Fears about Emerging Intelligent, Humanlike Machines* (Berlin: Springer Science & Business Media, 2009).

107   **through body language, tone, gesture, and facial expression:** Beatrice de Gelder, "Towards the Neurobiology of Emotional Body Language," *Nature Reviews Neuroscience* 7, no. 3 (2006): 242–49; Marius V. Peelen and Paul E. Downing, "The Neural Basis of Visual Body Perception," *Nature Reviews Neuroscience* 8, no. 8 (2007): 636–48; Annett Schirmer and Ralph Adolphs, "Emotion Perception from Face, Voice, and Touch: Comparisons and Convergence," *Trends in Cognitive Sciences* 21, no. 3 (2017): 216–28.

107   **her group has created a robotic chair:** Abhijeet Agnihotri, Alison Shutterly, Abrar Fallatah, Brian Layng, and Heather Knight, "ChairBot Café: Personality-Based Expressive Motion," in *Proceedings of the 2018 ACM/IEEE International Conference on Human-Robot Interaction Companion (HRI'18 Companion)* (2018): 14.

108   **a desk lamp that expresses emotion through motion:** Guy Hoffman, "Ensemble: Fluency and Embodiment for Robots Acting with Humans" (PhD diss., Massachusetts Institute of Technology, 2007).

109   **the "uncanny valley":** Masahiro Mori, "The Uncanny Valley: The Original Essay by Masahiro Mori," trans. Karl F. MacDorman and Norri Kageki, in *IEEE Spectrum*, June 12, 2012, https://spectrum.ieee.org/automaton/robotics /humanoids/the-uncanny-valley.

109   **Not everyone agrees with the hypothesis:** Angela Tinwell, Mark Grimshaw, and Andrew Williams, "The Uncanny Wall," *International Journal of Arts and Technology* 4, no. 3 (2011): 326–41.

109   **results have been mixed:** Jari Kätsyri, Klaus Förger, Meeri Mäkäräinen, and Tapio Takala, "A Review of Empirical Evidence on Different Uncanny Valley Hypotheses: Support for Perceptual Mismatch as One Road to the Valley of Eeriness," *Frontiers in Psychology* 6 (2015): 390; Karl F. MacDorman and Debaleena Chattopadhyay, "Reducing Consistency in Human Realism Increases the Uncanny Valley Effect; Increasing Category Uncertainty Does Not," *Cognition* 146 (2016): 190–205.

109   **One of the areas where it bears out in some form:** Kate Devlin, *Turned On: Science, Sex and Robots* (New York: Bloomsbury, 2018). ("Conflicting perceptual cues cause cognitive discomfort.")

110   **tricking human competition judges into thinking it was human 33 percent of the time:** Dante D'Orazio, "Computer Allegedly Passes Turing Test for
First Time by Convincing Judges It Is a 13-Year-Old Boy," *The Verge,* June 8,
2014, https://www.theverge.com/2014/6/8/5790936/computer-passes-turing
-test-for-first-time-by-convincing-judges-it-is.

110   **ELIZA:** Joseph Weizenbaum, *Computer Power and Human Reason: From
Judgment to Calculation* (New York: W. H. Freeman, 1976).

111   **In a 2018 survey of screen-rendered robot faces:** Alisa Kalegina, Grace
Schroeder, Aidan Allchin, Keara Berlin, and Maya Cakmak, "Characterizing
the Design Space of Rendered Robot Faces," in *Proceedings of the 2018 ACM/
IEEE International Conference on Human-Robot Interaction* (2018): 96–104.

111   **Communication through simple cues that don't require . . . is often more
effective:** Frédéric Kaplan, "Artificial Attachment: Will a Robot Ever Pass
Ainsworth's Strange Situation Test?," in *Proceedings of Humanoids* (2001):
125–32.

111   **randomness and unpredictable behavior can increase people's anthropomorphism:** Adam Waytz et al., "Making Sense by Making Sentient: Effec-
tance Motivation Increases Anthropomorphism," *Journal of Personality and
Social Psychology* 99, no. 3 (2010): 410.

111   **Research in the rapidly growing field of human-robot interaction:** Cory
D. Kidd and Cynthia Breazeal, "Comparison of Social Presence in Robots
and Animated Characters," *Interaction Journal Studies* (2005); Victoria
Groom, "What's the Best Role for a Robot? Cybernetic Models of Existing
and Proposed Human-Robot Interaction Structures," in *Proceedings of the
International Conference on Informatics in Control, Automation and Robot-
ics (ICINCO)* (2008): 323–28; Frank Hegel, Soren Krach, Tilo Kircher, Britta
Wrede, and Gerhard Sagerer, "Understanding Social Robots: A User Study on
Anthropomorphism," in *RO-MAN 2008—the 17th IEEE International Sympo-
sium on Robot and Human Interactive Communication* (2008): 574–79; Brian
R. Duffy, "Anthropomorphism and the Social Robot," *Robotics and Autono-
mous Systems* 42, nos. 3–4 (2003): 177–90.

5. (HU)MAN'S BEST FRIEND: THE HISTORY OF COMPANION ANIMALS

113   **Quote:** Isaac Asimov, *Robot Visions* (New York: Penguin Books, 1990).

113   **Dogs for Defense:** Susan Orlean, *Rin Tin Tin: The Life and Legend of the
World's Most Famous Dog* (London: Atlantic Books, 2012); "Animal Sacrifice," *This American Life,* November 30, 2012, https://www.thisamericanlife
.org/480/animal-sacrifice.

THE HISTORY OF PETS

114   **Rebecca the White House raccoon:** "Coolidge Gets a Raccoon," *New York
Times,* November 27, 1926, 4; "Coolidge 'Coon' Gets Ribbon and Is Now

Named Rebecca," *New York Times*, December 25, 1926, 2; Matthew Costello, "Raccoons at the White House," White House Historical Association, June 8, 2018, https://www.whitehousehistory.org/raccoons-at-the-white-house; "White House Raccoon Is AWOL Two Hours," *Indianapolis Star*, June 9, 1927.

115   **"a domesticated animal kept for pleasure rather than utility":** *Merriam-Webster* (2002).

115   **a painting from around 1580:** Maev Kennedy, "Elizabethan Portraits Offer Snapshot of Fashion for Exotic Pets," *Guardian*, August 20, 2013, http://www.theguardian.com/artanddesign/2013/aug/20/elizabethan-portraits-snapshot-fashion-exotic-pets.

116   **The ancient Greek word** *athurma*: Anthony L. Podberscek, Elizabeth S. Paul, and James A. Serpell, eds., *Companion Animals and Us: Exploring the Relationships between People and Pets* (Cambridge: Cambridge University Press, 2005).

116   **possibly beginning a transition to "dog" as early as 40,000 years ago:** Laura R. Botigué, Shiya Song, Amelie Scheu, Shyamalika Gopalan, Amanda L. Pendleton, Matthew Oetjens, Angela M. Taravella et al., "Ancient European Dog Genomes Reveal Continuity since the Early Neolithic," *Nature Communications* 8, no. 1 (2017): 1–11; see also Brian Handwerk, "How Accurate Is Alpha's Theory of Dog Domestication?," *Smithsonian Magazine*, August 15, 2018, https://www.smithsonianmag.com/science-nature/how-wolves-really-became-dogs-180970014/.

116   **we started to use dogs to help with hunting:** James A. Serpell, ed., *The Domestic Dog: Its Evolution, Behaviour and Interactions with People* (Cambridge: Cambridge University Press, 1995).

116   **coevolving as cooperative species:** Wolfgang M. Schleidt and Michael D. Shalter, "Co-evolution of Humans and Canids: An Alternative View of Dog Domestication; Homo Homini Lupus?," *Evolution and Cognition* 9, no. 1 (2003); Friederike Range and Zsófia Virányi, "Tracking the Evolutionary Origins of Dog-Human Cooperation: The "'Canine Cooperation Hypothesis.'" *Frontiers in Psychology* 5 (2015): 1582.

116   **dogs still have a variety of functions:** Stan Braude and Justin Gladman, "Out of Asia: An Allopatric Model for the Evolution of the Domestic Dog," *International Scholarly Research Notices* 2013 (2013); see also, for example, Carla Thomas, "Why China's Yulin Dog Meat Festival Won't Be Cancelled This Year After All," *Forbes*, June 15, 2017, https://www.forbes.com/sites/cfthomas/2017/06/15/why-chinas-yulin-dog-meat-festival-wont-be-cancelled-this-year-after-all/#fe104cb2a137.

116   **Dogs took on different roles across Indigenous cultures:** Marion Schwartz, *A History of Dogs in the Early Americas* (New Haven, CT: Yale University Press, 1998); Peter B. Gray and Sharon M. Young, "Human–Pet Dynamics in Cross-Cultural Perspective," *Anthrozoös* 24, no. 1 (2011): 17–30.

117   **Abuwtiyuw:** George A. Reisner, "The Dog Which Was Honored by the King of Upper and Lower Egypt," *Bulletin of the Museum of Fine Arts* 34, no. 206 (1936): 96–99.

117   **Pets in ancient Egypt:** Joshua J. Mark, "Pets in Ancient Egypt," *Ancient History Encyclopedia*, March 18, 2016, https://www.ancient.eu/article/875/pets-in-ancient-egypt/; Wim Van Neer, Veerle Linseele, Renée Friedman, and Bea De Cupere, "More Evidence for Cat Taming at the Predynastic Elite Cemetery of Hierakonpolis (Upper Egypt)," *Journal of Archaeological Science* 45 (2014): 103–11; Carlos A. Driscoll, Marilyn Menotti-Raymond, Alfred L. Roca, Karsten Hupe, Warren E. Johnson, Eli Geffen, Eric H. Harley et al., "The Near Eastern Origin of Cat Domestication," *Science* 317, no. 5837 (2007): 519–23; Jean-Denis Vigne, Jean Guilaine, Karyne Debue, Laurent Haye, and Patrice Gérard, "Early Taming of the Cat in Cyprus," *Science* 304, no. 5668 (2004): 259.

117   **Central Asian societies housed horses:** Juliet Clutton-Brock, *A Natural History of Domesticated Mammals* (Cambridge: Cambridge University Press, 1999).

117   **Some hunter-gatherer societies kept pets, such as monkeys:** Loretta A. Cormier, *Kinship with Monkeys: The Guajá Foragers of Eastern Amazonia* (New York: Columbia University Press, 2003); Stefan Seitz, "Game, Pets and Animal Husbandry among Penan and Punan Groups," in *Beyond the Green Myth: Borneo's Hunter-Gatherers in the Twenty-First Century*, ed. Peter Sercombe and Bernard Sellato (Copenhagen: NIAS Press, 2007), 177–91.

117   **same range of animals as pets that we do today:** Katherine C. Grier, *Pets in America: A History* (Chapel Hill: University of North Carolina Press, 2010).

117   **they didn't explode in popularity until after the 1914 Great Tokyo Exhibition:** Sheila Keenan, *Animals in the House: A History of Pets and People* (New York: Scholastic Nonfiction, 2007).

118   **12 percent of American households having fish of any sort at all:** Richard W. Bulliet, *Hunters, Herders, and Hamburgers: The Past and Future of Human-Animal Relationships* (New York: Columbia University Press, 2005).

118   **The East Asian, Aztec, and Western European elites loved small dogs:** Gray and Young, "Human–Pet Dynamics in Cross-Cultural Perspective," 2011.

118   **Chinese emperors would have anyone who stole one executed:** Michael Wines, "Once Banned, Dogs Reflect China's Rise," *New York Times*, October 24, 2010, https://www.nytimes.com/2010/10/25/world/asia/25dogs.html.

118   **capturing canaries in the Canary Islands and selling them to wealthy Europeans in the 1500s:** Keenan, *Animals in the House*.

118   **carrying his favorite dogs in a basket hanging around his neck:** Katharine MacDonogh, *Reigning Cats and Dogs: A History of Pets at Court Since the Renaissance* (London: Fourth Estate, 1999).

118   **negotiated its release through a proper prisoner of war exchange:** Keenan, *Animals in the House*.

118   **having clothes made with special pockets into which they could stuff
        their tiny dogs:** Alicia Ault, "Ask Smithsonian: When Did People Start
        Keeping Pets?," *Smithsonian Magazine*, September 28, 2016, https://www
        .smithsonianmag.com/smithsonian-institution/ask-smithsonian-when-did
        -people-start-keeping-pets-180960616/.

118   **they were commonly kidnapped for ransom or reward:** Zoë Crew, "A Look
        at Dog Theft in the Nineteenth-Century Slum," Exploring the Slum, Decem-
        ber 18, 2015, https://slumexplorers.wordpress.com/animals/zoe/.

119   **Beijing dog ban:** Wines, "Once Banned, Dogs Reflect China's Rise,"; Shelly
        Volsche, "Understanding Cross-Species Parenting: A Case for Pets as Chil-
        dren," in *Clinician's Guide to Treating Companion Animal Issues: Addressing
        Human-Animal Interaction*, ed. Lori Kogan and Christopher Blazina (Lon-
        don: Academic Press, 2018), 129–41.

119   **The dog trend in China follows those of other countries:** Volsche, "Under-
        standing Cross-Species Parenting"; James A. Serpell and Elizabeth S. Paul,
        "Pets in the Family: An Evolutionary Perspective," in *The Oxford Handbook
        of Evolutionary Family Psychology*, ed. Todd K. Shackelford and Catherine A.
        Salmon (New York: Oxford University Press, 2011), 297.

119   **that relationship is evolving, too:** Volsche, "Understanding Cross-Species
        Parenting."

119   **American Pet Products Association national pet owners survey:** "2019–
        2020 APPA National Pet Owners Survey," American Pet Products Associa-
        tion, accessed August 27, 2020, https://www.americanpetproducts.org/pubs
        _survey.asp.

119   **the bigger change has been in how much time, money, and emotional
        energy:** Jeffrey D. Green, Maureen A. Mathews, and Craig A. Foster,
        "Another Kind of 'Interpersonal' Relationship: Humans, Companion Ani-
        mals, and Attachment Theory," *Relationships and Psychology: A Practical
        Guide* (2009): 87–108.

119   **Now, in 2020, the industry is approaching a worth of $100 billion:** "Pet
        Industry Market Size & Ownership Statistics," American Pet Products Asso-
        ciation, accessed August 27, 2020, https://www.americanpetproducts.org/press
        _industrytrends.asp.

ANTHROPOMORPHISM (PROJECTING OURSELVES)

120   **cut and dye your fluffy friend to look like a panda or a Ninja Turtle:**
        Melinda Liu, "China's Dog Dyeing Craze: Once Shunned, Pet Pooches Now
        Embraced," *Daily Beast*, July 8, 2012, https://www.thedailybeast.com/chinas
        -dog-dyeing-craze-once-shunned-pet-pooches-now-embraced.

120   **a consistent increase in companion animal services that cater directly to
        people's anthropomorphism:** "Industry Statistics & Trends," American Pet
        Products Association; Nikolina M. Duvall Antonacopoulos and Timothy A.
        Pychyl, "The Possible Role of Companion-Animal Anthropomorphism and

Social Support in the Physical and Psychological Health of Dog Guardians," *Society & Animals* 18, no. 4 (2010): 379–95.

120   **An airline named Pet Airways:** Kathy McCabe, "Pet Travel: One Dog's Flight on Pet Airways," *HuffPost*, June 22, 2011, https://www.huffpost.com/entry/finney-flies-one-dogs-jou_b_881908.

121   **In 1890s Paris, people would dress up their pet dogs:** Hal Herzog, *Some We Love, Some We Hate, Some We Eat: Why It's So Hard to Think Straight about Animals* (New York: HarperCollins, 2010).

121   **show signs of excessive human pampering and feeding:** Michael Mac-Kinnon, "'Sick as a Dog': Zooarchaeological Evidence for Pet Dog Health and Welfare in the Roman World," *World Archaeology* 42, no. 2 (2010): 290–309.

121   **Kids' books increasingly placed animals at the center of the story:** Grier, *Pets in America*.

121   **the printed trade cards that people loved to collect included lots of pet animals:** Grier, *Pets in America*.

121   **William J. Long published multiple books about the animals:** William Joseph Long, *School of the Woods: Some Life Studies of Animal Instincts and Animal Training* (Toronto: Copp Clark, 1903); William Joseph Long, *A Little Brother to the Bear and Other Animal Studies* (Boston: Ginn, 1903); see also Michael Pettit, "The Problem of Raccoon Intelligence in Behaviourist America," *British Journal for the History of Science* (2010): 391–421.

121   **Clever Hans:** Oskar Pfungst, *Clever Hans (The Horse of Mr. Von Osten): A Contribution to Experimental Animal and Human Psychology* (New York: Holt, Rinehart and Winston, 1911).

121   **Dogs in the movies:** Kathryn Fuller-Seeley and Jeremy Groskopf, "The Dogs Who Saved Hollywood: Strongheart and Rin Tin Tin," in *Cinematic Canines: Dogs and Their Work in the Fiction Film*, ed. Adrienne L. McLean (New Brunswick, NJ: Rutgers University Press, 2014).

122   **"perfect" dog Lassie:** Patricia B. McConnell, *For the Love of a Dog: Understanding Emotion in You and Your Best Friend* (New York: Ballantine Books, 2009); Orlean, *Rin Tin Tin*; Ketzel Levine, "Lassie: The Perfect Dog Sets High Bar for Real Pups," NPR's *Morning Edition*, January 7, 2008, https://www.npr.org/templates/story/story.php?storyId=17894690.

123   **will falsely attribute similar types of mentalities to both dogs and children:** Jeffrey Lee Rasmussen and D. W. Rajecki, "Differences and Similarities in Humans' Perceptions of the Thinking and Feeling of a Dog and a Boy," *Society & Animals* 3, no. 2 (1995): 117–37.

123   **not a single one of the pets lifted a paw to save their human significant others:** Krista Macpherson and William A. Roberts, "Do Dogs (Canis Familiaris) Seek Help in an Emergency?," *Journal of Comparative Psychology* 120, no. 2 (2006): 113.

123   **a study showed that dog owners get it wrong:** Julie Hecht, Ádám Miklósi, and Márta Gácsi, "Behavioral Assessment and Owner Perceptions of

Behaviors Associated with Guilt in Dogs," *Applied Animal Behaviour Science* 139, no. 1–2 (2012): 134–42.

123   **it's nearly impossible for us to stop ourselves from anthropomorphizing animals:** James A. Serpell, "Anthropomorphism and Anthropomorphic Selection—Beyond the 'Cute Response,'" *Society & Animals* 11, no. 1 (2003): 83–100.

123   **is the main reason we have pets at all:** Steven Mithen, "The Prehistory of the Mind," *Cambridge Archaeological Journal* 7 (1997): 269; Serpell, "Beyond the 'Cute Response.'"

124   **noticed a sharp increase in dogs with heart diseases:** Linda Carroll, "FDA Names 16 Brands of Dog Food Linked to Canine Heart Disease," NBC News, July 1, 2019, https://www.nbcnews.com/health/health-news/fda-names-16 -brands-dog-food-linked-canine-heart-disease-n1025466.

124   **"project onto their pets whatever they think they need to eat themselves":** Amanda Mull, "How Americans Decided Dogs Can't Eat Grains," *Atlantic*, July 2, 2019, https://www.theatlantic.com/health/archive/2019/07/grain-free -dog-food-fda-warning/593167/.

BREEDING (DESIGN)

124   **bred in a university lab in Jerusalem:** Keenan, *Animals in the House.*

125   **Poodle fad:** Lisa Waller Rogers, "Doris Day Was Poodle-Icious," *Lisa's History Room*, February 10, 2010, https://lisawallerrogers.com/2010/02/10 /doris-day-was-poodle-icious/.

125   **less popular than breeds like retrievers and bulldogs:** "The Most Popular Dog Breeds of 2019," American Kennel Club, May 1, 2020, https://www.akc .org/expert-advice/dog-breeds/2020-popular-breeds-2019/.

125   **poodle-grooming documentary *Well Groomed*:** *Well Groomed*, written and directed by Rebecca Stern, Cattle Rat Productions in association with Spacestation, premiered December 11, 2019, on HBO, https://www.hbo.com /documentaries/well-groomed.

126   **growing to full size in four months and ballooning up three times larger than wild turkeys:** "Large Breed Turkey Special Care Considerations," Open Sanctuary Project, accessed August 30, 2020, https://opensanctuary.org /article/large-breed-turkey-special-care-considerations/.

126   **bred sled dogs in remote Siberia as early as 9,000 years ago:** David Grimm, "Earliest Evidence for Dog Breeding Found on Remote Siberian Island," *Science*, May 26, 2017, https://www.sciencemag.org/news/2017/05/earliest -evidence-dog-breeding-found-remote-siberian-island.

126   **albino and other unusually colored mice in the 1700s:** Keenan, *Animals in the House.*

126   **cat shows exhibiting a wide variety of different felines starting as early as 1871 in London:** Kirsten Fawcett, "A Brief, Fur-Filled History of Cat Shows,"

Mental Floss, December 29, 2015, https://www.mentalfloss.com/article/72203
/brief-fur-filled-history-cat-shows.

126   **cultivating more exotic-seeming breeds like Siamese cats was all the rage:**
       Grier, *Pets in America.*

126   **the wolf genes that we've been able to shape into so many different canine
       skills and appearances:** Grier, *Pets in America*; Herzog, *Some We Love.*

126   **First bred by Australian Wally Conron in 1989:** Kashmira Gander, "Labra-
       doodle 'Inventor' Wally Conron Admits He's 'Done a Lot of Damage' Ahead
       of US Westminster Kennel Club Dog Show," *Independent*, February 7, 2014,
       http://www.independent.co.uk/news/world/americas/labradoodle-inventor
       -wally-conron-admits-hes-done-a-lot-of-damage-ahead-of-us-westminster
       -kennel-9115851.html; Rory Sullivan, "Labradoodle Creator Says He Regrets
       'Frankenstein's Monster,'" CNN, September 26, 2019, https://www.cnn.com
       /2019/09/26/world/labradoodle-creator-frankenstein-monster-scli-intl/index
       .html.

127   **are about as aggressive as pit bulls and rottweilers:** Deborah L. Duffy,
       Yuying Hsu, and James A. Serpell, "Breed Differences in Canine Aggression,"
       Applied Animal Behaviour Science 114, no. 3–4 (2008): 441–60.

127   **the "cute response":** Marta Borgi and Francesca Cirulli, "Pet Face: Mecha-
       nisms Underlying Human-Animal Relationships," *Frontiers in Psychology* 7
       (2016): 298.

127   **Breeding health problems:** Serpell, "Beyond the 'Cute Response,'" 83–100;
       Grier, *Pets in America*; The Orthopedic Foundation for Animals was founded
       in 1966, see "History," Orthopedic Foundation for Animals, accessed August
       29, 2020, https://www.ofa.org/about/history.

128   **"like modern tomatoes, a triumph of style over substance":** Herzog, *Some
       We Love.*

128   **Ira Glass's dog:** "Animal Sacrifice," *This American Life*, November 30, 2012,
       https://www.thisamericanlife.org/480/animal-sacrifice.

COMPANIONS IN HEALTH AND EDUCATION

128   **Boris Levinson and Jingles:** Jill Lenk Schilp, *Dogs in Health Care: Pioneer-
       ing Animal-Human Partnerships* (Jefferson, NC: McFarland, 2019); Stanley
       Coren, "How Therapy Dogs Almost Never Came to Exist," *Psychology Today*,
       February 11, 2013; Gerald P. Mallon, "A Generous Spirit: The Work and Life
       of Boris Levinson," *Anthrozoös* 7, no. 4 (1994): 224–31.

129   **Freud had documented effects similar to what Levinson had independently
       discovered:** Coren, "How Therapy Dogs Almost Never Came to Exist."

129   **the ancient Greeks used horses to improve the moods of sick patients:**
       Mira Lessick, Robyn Shinaver, Kimberly M. Post, Jennifer E. Rivera, and
       Betty Lemon, "Therapeutic Horseback Riding," *AWHONN Lifelines* 8, no. 1
       (2004): 46–53.

130   Quakers believed that the animals could help the patients with socializa-
        tion and behavioral management: Samuel Tuke, *Description of the Retreat:
        An Institution Near York, for Insane Persons of the Society of Friends* (Oxford:
        W. Alexander, 1813).

130   Florence Nightingale, touted the benefits of keeping birds: Florence
        Nightingale, *Notes on Nursing: What It Is and What It Is Not* (London: Har-
        rison and Sons, 1859).

130   The Bethlem Hospital in England soon introduced a variety of animals:
        Patricia H. Allderidge, "A Cat, Surpassing in Beauty, and Other Therapeutic
        Animals," *Psychiatric Bulletin* 15, no. 12 (1991): 759–62.

130   in Germany, the use of animals was expanded to help with other disorders
        like epilepsy: Jonica Newby, *The Animal Attraction: Humans and Their Ani-
        mal Companions* (Sydney: ABC Books, 1997).

130   a military program for veterans in the 1940s: DOD Human-Animal Bond
        Principles and Guidelines, June 16, 2003, available at https://archive.org
        /stream/ost-military-medical-tbmed4/tbmed4_djvu.txt; Katherine Connor
        and Julie Miller, "Animal-Assisted Therapy: An In-Depth Look," *Dimensions
        of Critical Care Nursing* 19, no. 3 (2000): 20.

130   The San Jose airport introduced a therapy dog program after 9/11: "Ther-
        apy Animals," San Jose International, accessed August 29, 2020, https://www
        .flysanjose.com/therapy-animals.

130   At MIT, our stressed-out students can relax by cuddling therapy dogs:
        Sarah Goodman, "MIT Puppy Lab to Open during National Mental Health
        Awareness Month," *MIT News*, May 4, 2016, https://news.mit.edu/2016
        /mit-puppy-lab-open-during-national-mental-health-awareness-month
        -0504.

130   or just find them a little "too much": Jennifer A. Kingson, "The Llama as Ther-
        apist," *New York Times*, November 14, 2019, https://www.nytimes.com/2019
        /11/14/well/live/llama-therapy-nursing-homes-elderly-seniors-hospitals-pet
        -therapy.html.

130   "Caesar the No Drama Llama": Molly O'Brien, "Therapy Llama 'Caesar the
        No Drama Llama' Calms Tensions at Protests," *Washington Post*, August 5,
        2020, https://www.washingtonpost.com/lifestyle/2020/08/05/therapy-llama
        -caesar-no-drama-llama-diffuses-tension-protests/.

130   "that improves the physical, social, and emotional lives of those we serve":
        "Who We Are," Pet Partners, accessed August 29, 2020, https://petpartners
        .org/about-us/who-we-are/.

130   John Bradshaw . . . cautions against over-interpreting positive results:
        John Bradshaw, "The Truth about Pet Therapy: Animals Probably Don't Help
        Health Problems After All," inews.co.uk, September 27, 2017, https://inews
        .co.uk/news/long-reads/pet-therapy-health-treatments-john-bradshaw
        -93604.

131   The efficacy of animal therapy doesn't always match what we want to believe:
        Hal Herzog, "Does Dolphin Therapy Work?," *Psychology Today*, October 9,

2011, http://www.psychologytoday.com/blog/animals-and-us/201110/does-dolphin-therapy-work.

131   **Dolphin-assisted therapy:** Alexis McKinney, Dan Dustin, and Robert Wolff, "The Promise of Dolphin-Assisted Therapy," *Parks & Recreation* 36, no. 5 (2001): 46–50; Britta L. Fiksdal, Daniel Houlihan, and Aaron C. Barnes, "Dolphin-Assisted Therapy: Claims versus Evidence," *Autism Research and Treatment* 2012 (2012); Lori Marino and Scott O. Lilienfeld, "Dolphin-Assisted Therapy: Flawed Data, Flawed Conclusions," *Anthrozoös* 11, no. 4 (1998): 194–200; Lori Marino and Scott O. Lilienfeld, "Dolphin-Assisted Therapy: More Flawed Data and More Flawed Conclusions," *Anthrozoös* 20, no. 3 (2007): 239–49.

131   **Even these are sometimes contested:** See, for example, Ruth A. Parslow, Anthony F. Jorm, Helen Christensen, Bryan Rodgers, and Patricia Jacomb, "Pet Ownership and Health in Older Adults: Findings from a Survey of 2,551 Community-Based Australians Aged 60–64," *Gerontology* 51, no. 1 (2005): 40–47; Nancy A. Pachana, Jessica H. Ford, Brooke Andrew, and Annette J. Dobson, "Relations between Companion Animals and Self-Reported Health in Older Women: Cause, Effect or Artifact?," *International Journal of Behavioral Medicine* 12, no. 2 (2005): 103.

131   **whether healthy people are simply more likely to get a pet:** Bradshaw, "Truth about Pet Therapy."

131   **a 2018 study of elderly Japanese community dwellers:** Yu Taniguchi, Satoshi Seino, Mariko Nishi, Yui Tomine, Izumi Tanaka, Yuri Yokoyama, Hidenori Amano, Akihiko Kitamura, and Shoji Shinkai, "Physical, Social, and Psychological Characteristics of Community-Dwelling Elderly Japanese Dog and Cat Owners," *PLOS ONE* 13, no. 11 (2018): e0206399.

131   **decline in people's activity and health after getting a pet:** Leena K. Koivusilta and Ansa Ojanlatva, "To Have or Not to Have a Pet for Better Health?," *PLOS ONE* 1, no. 1 (2006): e109.

132   **the belief that pets promote people's health and well-being is not scientifically unfounded:** Emily K. Crawford, Nancy L. Worsham, and Elizabeth R. Swinehart, "Benefits Derived from Companion Animals, and the Use of the Term 'Attachment,'" *Anthrozoös* 19, no. 2 (2006): 98–112; Ariann E. Robino, "The Human-Animal Bond and Attachment in Animal-Assisted Interventions in Counseling" (PhD diss., Virginia Tech, 2019); Allen R. McConnell, Christina M. Brown, Tonya M. Shoda, Laura E. Stayton, and Colleen E. Martin, "Friends with Benefits: On the Positive Consequences of Pet Ownership," *Journal of Personality and Social Psychology* 101, no. 6 (2011): 1239; Sandra B. Barker, Janet S. Knisely, Nancy L. McCain, Christine M. Schubert, and Anand K. Pandurangi, "Exploratory Study of Stress-Buffering Response Patterns from Interaction with a Therapy Dog," *Anthrozoös* 23, no. 1 (2010): 79–91; Carl J. Charnetski, Sandra Riggers, and Francis X. Brennan, "Effect of Petting a Dog on Immune System Function," *Psychological Reports* 95, no. 3, suppl. (2004): 1087–91; Christine Olsen, Ingeborg Pedersen, Astrid Bergland,

Marie-José Enders-Slegers, Grete Patil, and Camilla Ihlebæk, "Effect of Animal-Assisted Interventions on Depression, Agitation and Quality of Life in Nursing Home Residents Suffering from Cognitive Impairment or Dementia: A Cluster Randomized Controlled Trial," *International Journal of Geriatric Psychiatry* 31, no. 12 (2016): 1312–21; Lahna Bradley and Pauleen C. Bennett, "Companion-Animals' Effectiveness in Managing Chronic Pain in Adult Community Members," *Anthrozoös* 28, no. 4 (2015): 635–47; Eija Bergroth, Sami Remes, Juha Pekkanen, Timo Kauppila, Gisela Büchele, and Leea Keski-Nisula, "Respiratory Tract Illnesses during the First Year of Life: Effect of Dog and Cat Contacts," *Pediatrics* 130, no. 2 (2012): 211–20; Mwenya Mubanga, Liisa Byberg, Christoph Nowak, Agneta Egenvall, Patrik K. Magnusson, Erik Ingelsson, and Tove Fall, "Dog Ownership and the Risk of Cardiovascular Disease and Death—a Nationwide Cohort Study," *Scientific Reports* 7, no. 1 (2017): 1–9; Adnan I. Qureshi, Muhammad Zeeshan Memon, Gabriela Vazquez, and M. Fareed K. Suri, "Cat Ownership and the Risk of Fatal Cardiovascular Diseases. Results from the Second National Health and Nutrition Examination Study Mortality Follow-up Study," *Journal of Vascular and Interventional Neurology* 2, no. 1 (2009): 132; Nikolina M. Duvall Antonacopoulos and Timothy A. Pychyl, "An Examination of the Potential Role of Pet Ownership, Human Social Support and Pet Attachment in the Psychological Health of Individuals Living Alone," *Anthrozoös* 23, no. 1 (2010): 37–54; Miho Nagasawa and Mitsuaki Ohta, "The Influence of Dog Ownership in Childhood on the Sociality of Elderly Japanese Men," *Animal Science Journal* 81, no. 3 (2010): 377–83; Parminder Raina, David Waltner-Toews, Brenda Bonnett, Christel Woodward, and Tom Abernathy, "Influence of Companion Animals on the Physical and Psychological Health of Older People: An Analysis of a One-Year Longitudinal Study," *Journal of the American Geriatrics Society* 47, no. 3 (1999): 323–29; Green, Mathews, and Foster, "Another Kind of 'Interpersonal' Relationship," 87–108.

132   **thanks to the recently lifted dog bans, researchers were able to conduct a natural experiment:** Bruce Headey, Fu Na, and Richard Zheng, "Pet Dogs Benefit Owners' Health: A 'Natural Experiment'in China," *Social Indicators Research* 87, no. 3 (2008): 481–93.

132   **dog owners may walk more because Fido needs to go out:** Green, Mathews, and Foster, "Another Kind of 'Interpersonal' Relationship."

132   **the 2018 Japanese community study suggested that the social benefits were due to more opportunities:** Taniguchi et al., "Elderly Japanese Dog and Cat Owners."

132   **In a study of 3,465 prospective dog adopters:** Lauren Powell, Debbie Chia, Paul McGreevy, Anthony L. Podberscek, Kate M. Edwards, Brendon Neilly, Adam J. Guastella, Vanessa Lee, and Emmanuel Stamatakis, "Expectations for Dog Ownership: Perceived Physical, Mental and Psychosocial Health Consequences among Prospective Adopters," *PLOS ONE* 13, no. 7 (2018): e0200276.

133   **"Some are our best friends":** Grier, *Pets in America.*

133   **"friend, servant, admirer, confidante" to "slave, scapegoat, a mirror of trust or a defender":** Cited in Schilp, *Dogs in Health Care.*

133   **animals can be a symbol of qualities, like bravery or strength:** Susan D. Greenbaum, "Introduction to Working with Animal Assisted Crisis Response Animal Handler Teams," *International Journal of Emergency Mental Health* 8, no. 1 (2006): 49–63.

133   **A 2007 meta-study of forty-nine different studies:** Janelle Nimer and Brad Lundahl, "Animal-Assisted Therapy: A Meta-Analysis," *Anthrozoös* 20, no. 3 (2007): 225–38.

133   **"When a child needs to love safely . . . without losing face, a dog can provide this":** Boris M. Levinson, "The Dog as a 'Co-Therapist,'" *Mental Hygiene* 46 (1962): 59–65.

133   **When researcher Karen Allen and her colleagues had dog owners:** Karen M. Allen, Jim Blascovich, Joe Tomaka, and Robert M. Kelsey, "Presence of Human Friends and Pet Dogs as Moderators of Autonomic Responses to Stress in Women," *Journal of Personality and Social Psychology* 61, no. 4 (1991): 582.

134   **animals to help with academic learning and reading:** Matthew Philips, "Education: A Reader's Best Friend," *Newsweek*, December 3, 2006, https:// www.newsweek.com/education-readers-best-friend-105295; Juli Fraga, "How Reading Aloud to Therapy Dogs Can Help Struggling Kids," *MindShift* (blog), KQED, February 13, 2017, https://www.kqed.org/mindshift/47522/how -reading-aloud-to-therapy-dogs-can-help-struggling-kids.

134   **having your child read to a stuffed animal:** Fraga, "Reading Aloud."

6. A NEW CATEGORY OF RELATIONSHIP

135   **Quote:** Donna Haraway, *The Haraway Reader* (New York: Routledge, 2004).

135   **Gayndah:** Gayndah destination information, official website for Queensland, Australia, accessed August 29, 2020, https://www.queensland.com/us /en/places-to-see/destination-information/p-56b25e2c2cbcbe7073ad7a02 -gayndah.html; "2016 Census QuickStats: Gayndah," Australian Bureau of Statistics, accessed August 29, 2020, https://quickstats.censusdata.abs.gov .au/census_services/getproduct/census/2016/quickstat/SSC31118.

135   **PARO at Gunther Village:** "PARO Therapeutic Robot," PARO Robots, accessed August 29, 2020, http://www.parorobots.com/index.asp; Jess Lodge, "Paro the Robotic Seal Provides Unconditional Love for Aged Care Residents," ABC News Australia, July 15, 2018, https://www.abc.net.au/news/2018-07-16 /paro-the-robotic-seal-brings-comfort-to-aged-care-residents/9947946.

136   **Over a decade of research supports the caretakers' and families' intuition:** See, for example, Hayley Robinson, Bruce A. MacDonald, Ngaire Kerse, and Elizabeth Broadbent, "Suitability of Healthcare Robots for a Dementia Unit and Suggested Improvements," *Journal of the American Medical Directors*

Association 14, no. 1 (2013): 34–40; Kazuyoshi Wada and Takanori Shibata, "Living with Seal Robots—Its Sociopsychological and Physiological Influences on the Elderly at a Care House," *IEEE Transactions on Robotics* 23, no. 5 (2007): 972–80; Nina Jøranson, Ingeborg Pedersen, Anne Marie Mork Rokstad, and Camilla Ihlebæk, "Effects on Symptoms of Agitation and Depression in Persons with Dementia Participating in Robot-Assisted Activity: A Cluster-Randomized Controlled Trial," *Journal of the American Medical Directors Association* 16, no. 10 (2015): 867–73; Andrew Griffiths, "How Paro the Robot Seal Is Being Used to Help UK Dementia Patients," *Guardian*, July 8, 2014, https://www.theguardian.com/society/2014/jul/08/paro-robot-seal-dementia -patients-nhs-japan; Roger Bemelmans, Gert Jan Gelderblom, Pieter Jonker, and Luc de Witte, "Effectiveness of Robot Paro in Intramural Psychogeriatric Care: A Multicenter Quasi-Experimental Study," *Journal of the American Medical Directors Association* 16, no. 11 (2015): 946–50; S. M. Chang and H. C. Sung, "The Effectiveness of Paro Robot Therapy on Mood of Older Adults: A Systematic Review," *International Journal of Evidence-Based Healthcare* 11, no. 3 (2013): 216; Cory D. Kidd, Will Taggart, and Sherry Turkle, "A Sociable Robot to Encourage Social Interaction among the Elderly," in *Proceedings of the 2006 IEEE International Conference on Robotics and Automation, ICRA 2006* (2006): 3972–76.

SUPPLEMENT VERSUS REPLACEMENT

138 **For some . . . the thought of people bonding with PARO is worrisome:** Sherry Turkle, *Alone Together: Why We Expect More from Technology and Less from Each Other* (London: Hachette UK, 2017); Sherry Turkle, "A Nascent Robotics Culture: New Complicities for Companionship," *American Association for Artificial Intelligence Technical Report Series AAAI* (2006).

139 **Relationships with animals, they said, can be unhealthy, even pathological:** James A. Serpell, "People in Disguise: Anthropomorphism and the Human-Pet Relationship," in *Thinking with Animals: New Perspectives on Anthropomorphism*, ed. Lorraine Daston and Gregg Mitman (New York: Columbia University Press, 2005): 121–36; Edward K. Rynearson, "Humans and Pets and Attachment," *British Journal of Psychiatry* 133, no. 6 (1978): 550–55; Adrian Franklin, *Animals and Modern Cultures: A Sociology of Human-Animal Relations in Modernity* (London: Sage, 1999); see also Ariann E. Robino, "The Human-Animal Bond and Attachment in Animal-Assisted Interventions in Counseling" (PhD diss., Virginia Tech, 2019).

140 **suggest that separated, unpartnered, or childless people are more likely to use pets as substitutes:** Nikolina M. Duvall Antonacopoulos and Timothy A. Pychyl, "The Possible Role of Companion-Animal Anthropomorphism and Social Support in the Physical and Psychological Health of Dog Guardians," *Society & Animals* 18, no. 4 (2010): 379–95.

140   **Child-free pet owners don't confuse their animals for babies:** Shelly
        Volsche, "Understanding Cross-Species Parenting: A Case for Pets as Chil-
        dren," in *Clinician's Guide to Treating Companion Animal Issues: Addressing
        Human-Animal Interaction*, ed. Lori Kogan and Christopher Blazina (Lon-
        don: Academic Press, 2018), 129–41.

140   **A study conducted by researchers at UCLA found no difference:** Chris-
        tine E. Parsons, Richard T. LeBeau, Morten L. Kringelbach, and Katherine S.
        Young, "Pawsitively Sad: Pet-Owners Are More Sensitive to Negative Emo-
        tion in Animal Distress Vocalizations," *Royal Society Open Science* 6, no. 8
        (2019): 181555; see also F. Solmi, J. F. Hayes, G. Lewis, and J. B. Kirkbride,
        "Curiosity Killed the Cat: No Evidence of an Association between Cat Own-
        ership and Psychotic Symptoms at Ages 13 and 18 Years in a UK General
        Population Cohort," *Psychological Medicine* 47, no. 9 (2017): 1659–67.

140   **"pet parenting" . . . isn't particularly common:** Jean E. Veevers, *Childless
        by Choice* (Oxford: Butterworth-Heinemann, 1980).

141   **the nearly 85 million households that have a pet in America:** "Pet Indus-
        try Market Size & Ownership Statistics," American Pet Products Associa-
        tion, accessed August 27, 2020, https://www.americanpetproducts.org/press
        _industrytrends.asp; Serpell, "People in Disguise."

141   **senior citizens report in surveys that their dogs are their only friends:**
        See, for example, Peter O. Peretti, "Elderly-Animal Friendship Bonds," *Social
        Behavior and Personality: An International Journal* 18, no. 1 (1990): 151–56.

NEW COMPANIONS IN HEALTH AND EDUCATION

142   **Huggable:** Deirdre E. Logan, Cynthia Breazeal, Matthew S. Goodwin,
        Sooyeon Jeong, Brianna O'Connell, Duncan Smith-Freedman, James Heath-
        ers, and Peter Weinstock, "Social Robots for Hospitalized Children," *Pediat-
        rics* 144, no. 1 (2019): e20181511; Rob Matheson, "Study: Social Robots Can
        Benefit Hospitalized Children," *MIT News*, accessed August 29, 2020, https://
        news.mit.edu/2019/social-robots-benefit-sick-children-0626.

142   **adjust their responses according to the child's learning needs:** Gosia Glinska,
        "The Rise of Social Robots: How AI Can Help Us Flourish," Darden Ideas to
        Action, January 7, 2020, https://ideas.darden.virginia.edu/rise-of-social-robots.

142   **physical interaction that works so much better than a screen:** Evan Acker-
        man, "MIT's DragonBot Evolving to Better Teach Kids," *IEEE Spectrum*, March
        16, 2015, https://spectrum.ieee.org/automaton/robotics/artificial-intelligence
        /mit-dragonbot-evolving-to-better-teach-kids; for a study, see, for example,
        W. Bradley Knox, Samuel Spaulding, and Cynthia Breazeal, "Learning Social
        Interaction from the Wizard: A Proposal," in *Workshops at the Twenty-Eighth
        AAAI Conference on Artificial Intelligence* (2014).

142   **showing promise at getting kids to work hard:** "Robots That Teach Us about
        Ourselves," *Yale Engineering*, 2014–2015, http://yaleseas.com/magazine/Yale
        -Engineering-Magazine-2014-15.pdf.

143   **When a group of Yale researchers asked teachers:** Daniel Leyzberg, Aditi
        Ramachandran, and Brian Scassellati, "The Effect of Personalization in Longer-
        Term Robot Tutoring," *ACM Transactions on Human-Robot Interaction
        (THRI)* 7, no. 3 (2018): 1–19; see also Brian Scassellati, "Mapping the Frontier
        between Man and Machine" (lecture, Yale University, New Haven, CT, Sep-
        tember 4, 2019), https://whc.yale.edu/videos/mapping-frontier-between-man
        -and-machine.

144   **The above study, designed by his student Eli Kim:** Elizabeth S. Kim, Lauren
        D. Berkovits, Emily P. Bernier, Dan Leyzberg, Frederick Shic, Rhea Paul, and
        Brian Scassellati, "Social Robots as Embedded Reinforcers of Social Behav-
        ior in Children with Autism," *Journal of Autism and Developmental Disorders*
        43, no. 5 (2013): 1038–49.

144   **For nearly two decades, they've looked at using a variety of different
        robots:** "Yale University Social Robotics Lab," accessed August 30, 2020, https://
        scazlab.yale.edu/.

145   **Scaz, like Levinson before him, has a hunch:** William Weir, "Robots Help
        Children with Autism Improve Social Skills," *YaleNews*, August 22, 2018,
        https://news.yale.edu/2018/08/22/robots-help-children-autism-improve
        -social-skills.

145   **working closely . . . to test these new methods of engaging children:**
        Joshua J. Diehl, Lauren M. Schmitt, Michael Villano, and Charles R. Crowell,
        "The Clinical Use of Robots for Individuals with Autism Spectrum Disor-
        ders: A Critical Review," *Research in Autism Spectrum Disorders* 6, no. 1
        (2012): 249–62; Daniel J. Ricks and Mark B. Colton, "Trends and Consider-
        ations in Robot-Assisted Autism Therapy," in *2010 IEEE International Con-
        ference on Robotics and Automation* (2010): 4354–59; Brian Scassellati, "How
        Social Robots Will Help Us to Diagnose, Treat, and Understand Autism," in
        Robotics Research (Berlin: Springer, 2007), 552–63.

145   **Scaz recently led a groundbreaking study:** Brian Scassellati, Laura Boccan-
        fuso, Chien-Ming Huang, Marilena Mademtzi, Meiying Qin, Nicole Salomons,
        Pamela Ventola, and Frederick Shic, "Improving Social Skills in Children with
        ASD Using a Long-Term, In-Home Social Robot," *Science Robotics* 3, no. 21
        (2018); Weir, "Robots Help Children with Autism Improve Social Skills."

146   **without sufficient evidence that they're effective:** Mary E. McDonald,
        Darra Pace, Elfreda Blue, and Diane Schwartz, "Critical Issues in Causation
        and Treatment of Autism: Why Fads Continue to Flourish," *Child & Family
        Behavior Therapy* 34, no. 4 (2012): 290–304.

146   **a survey of 420 people showed that:** Mark Coeckelbergh, Cristina Pop,
        Ramona Simut, Andreea Peca, Sebastian Pintea, Daniel David, and Bram
        Vanderborght, "A Survey of Expectations about the Role of Robots in Robot-
        Assisted Therapy for Children with ASD: Ethical Acceptability, Trust, Socia-
        bility, Appearance, and Attachment," *Science and Engineering Ethics* 22, no. 1
        (2016): 47–65.

146   **Levinson: Animals could help expand treatment to more children who
        needed it:** Jill Lenk Schilp, *Dogs in Health Care: Pioneering Animal-Human
        Partnerships* (Jefferson, NC: McFarland, 2019).

ROBOTS VERSUS ANIMALS

147   **most animal "therapy" in elder care facilities is in the form of visitation:**
        Katherine Connor and Julie Miller, "Animal-Assisted Therapy: An In-Depth
        Look," *Dimensions of Critical Care Nursing* 19, no. 3 (2000): 20.

147   **as happened with two llamas in an Arizona assisted-living home:** Mark
        Berman, Elahe Izadi, Abby Ohlheiser, Tim Herrera, Sarah Larimer, and
        Christopher Ingraham, "BREAKING: Two Llamas Were on the Run Today in
        Arizona," *Washington Post*, February 26, 2015, https://www.washingtonpost
        .com/news/post-nation/wp/2015/02/26/breaking-two-llamas-are-on-the
        -run/.

147   **researchers in HRI have been exploring using robot-assisted therapy:**
        See, for example, Walter Dan Stiehl, Jeff Lieberman, Cynthia Breazeal, Louis
        Basel, Levi Lalla, and Michael Wolf, "Design of a Therapeutic Robotic Com-
        panion for Relational, Affective Touch," in *ROMAN 2005. IEEE Interna-
        tional Workshop on Robot and Human Interactive Communication* (2005):
        408–15; Emily C. Collins, "Drawing Parallels in Human–Other Interactions:
        A Trans-Disciplinary Approach to Developing Human–Robot Interaction
        Methodologies," *Philosophical Transactions of the Royal Society B* 374, no.
        1771 (2019): 20180433; Marian R. Banks, Lisa M. Willoughby, and William
        A. Banks, "Animal-Assisted Therapy and Loneliness in Nursing Homes: Use
        of Robotic versus Living Dogs," *Journal of the American Medical Directors
        Association* 9, no. 3 (2008): 173–77. (See also PARO research above.)

147   ***To Siri with Love:*** Judith Newman, *To Siri with Love: A Mother, Her Autistic
        Son, and the Kindness of Machines* (London: Hachette UK, 2017).

148   **Ellie is a robot head that makes mistakes:** Scassellati, "Mapping the Fron-
        tiers between Man and Machine."

RECIPROCITY

149   **1.5 million in the United States alone:** "Pet Statistics," ASPCA, accessed
        August 30, 2020, https://www.aspca.org/animal-homelessness/shelter-intake
        -and-surrender/pet-statistics.

149   **Researchers think we use them to create connection and belonging:** Wendi
        L. Gardner, Cynthia L. Pickett, and Megan Knowles, "Social Snacking and
        Shielding: Using Social Symbols, Selves, and Surrogates in the Service of
        Belonging Needs," in *The Social Outcast: Ostracism, Social Exclusion, Rejec-
        tion, and Bullying*, ed. Kipling D. Williams, Joseph P. Forgas, and William von
        Hippel (New York: Psychology Press, 2005): 227–42; see also Jeffrey D. Green,

Maureen A. Mathews, and Craig A. Foster, "Another Kind of 'Interpersonal' Relationship: Humans, Companion Animals, and Attachment Theory," *Relationships and Psychology: A Practical Guide* (2009): 87–108.

149   **Studies show that even just writing about or viewing a picture of a favorite TV show character:** Megan Knowles and Wendi L. Gardner, "'I'll Be There for You . . .': Favorite Television Characters as Social Surrogates" (paper presented at the Society for Personality and Social Psychology conference, Albuquerque, NM, February 2008).

149   **Connections to God or nature as attachment figures can make people feel:** Lee A. Kirkpatrick, "God as a Substitute Attachment Figure: A Longitudinal Study of Adult Attachment Style and Religious Change in College Students," *Personality and Social Psychology Bulletin* 24, no. 9 (1998): 961–73; Green, Mathews, and Foster, "Another Kind of 'Interpersonal' Relationship," 87–108.

150   **extend this view to our relationships with machines:** Donna Haraway, *The Companion Species Manifesto: Dogs, People, and Significant Otherness*, vol. 1 (Chicago: Prickly Paradigm Press, 2003).

150   **"If Pluto is Mickey Mouse's dog, then what on earth is Goofy?":** Susan McHugh, *Dog* (London: Reaktion Books, 2004).

BETTER TO HAVE LOVED AND LOST

150   **In 2007, the *Washington Post* reported:** Joel Garreau, "Bots on the Ground," *Washington Post*, May 6, 2007, http://www.washingtonpost.com/wp-dyn/content/article/2007/05/05/AR2007050501009.html.

151   **One day, iRobot received a letter from a US Navy chief petty officer:** P. W. Singer, *Wired for War: The Robotics Revolution and Conflict in the 21st Century* (New York: Penguin, 2009).

152   **emotional bonds with bomb disposal robots:** Julie Carpenter, *Culture and Human-Robot Interaction in Militarized Spaces: A War Story* (London: Routledge, 2016); Megan Garber, "Funerals for Fallen Robots," *Atlantic*, September 20, 2013, https://www.theatlantic.com/technology/archive/2013/09/funerals-for-fallen-robots/279861/; Garreau, "Bots on the Ground."

152   **P. W. Singer . . . describes how:** Singer, *Wired for War*.

153   **Horses in the Canadian Expeditionary Force:** Andrew McEwen, "'He Took Care of Me': The Human-Animal Bond in Canada's Great War," in *The Historical Animal*, ed. Susan Nance (Syracuse, NY: Syracuse University Press, 2015), 272–88.

153   **"The army was bad enough without having a horse to babysit":** James Robert Johnston, *Riding into War: The Memoir of a Horse Transport Driver, 1916–1919* (New Brunswick, Canada: Goose Lane Editions and New Brunswick Military Heritage Project Series, 2004).

154   **a brown bear named Wojtek that was raised by the Polish army:** Krystyna Mikula-Deegan, *Private Wojtek: Soldier Bear* (Kibworth Harcourt, UK: Troubador, 2011).

154 **In 2000, the US finally made it legal for the handlers to adopt:** "Robby's Law," H.R. 5314.

154 **"without the conversation returning to the subject of war":** Perry R. Chumley, "Historical Perspectives of the Human-Animal Bond within the Department of Defense," *US Army Medical Department Journal* (2012): 18–21.

154 **According to crisis response teams, animals can help:** Jan Shubert, "Therapy Dogs and Stress Management Assistance during Disasters," *US Army Medical Department Journal* (2012): 74–79.

154 **Roboticists talk about being distracted or even "disturbed":** Matthias Scheutz, "The Inherent Dangers of Unidirectional Emotional Bonds between Humans and Social Robots," in *Robot Ethics: The Ethical and Social Implications of Robotics*, ed. Patrick Lin, Keith Abney, and George A. Bekey (Cambridge, MA: MIT Press, 2011), 205.

155 **"we are programmed by biology such that we will necessarily be affected":** Joanna Bryson, "Is There a Human-AI Relationship?," September 25, 2019, https://web.archive.org/web/20200724112300/https://joanna-bryson.blogspot.com/2019/09/is-there-human-ai-relationship.html.

7. THE REAL ISSUES WITH ROBOT COMPANIONSHIP

156 **Quote:** Langdon Winner, "Do Artifacts Have Politics?," *Daedalus* 109, no. 1 (1980): 121–36.

156 **Maddy and Jibo:** "Death Online: Mourning a Robot," *Why'd You Push That Button?* (podcast), Vox Media, 2019, https://podcasts.apple.com/us/podcast/death-online-mourning-a-robot/id1295289748?i=1000442026036; Ashley Carman, "They Welcomed a Robot into Their Family, Now They're Mourning Its Death," *The Verge*, June 19, 2019, https://www.theverge.com/2019/6/19/18682780/jibo-death-server-update-social-robot-mourning.

COERCION

157 **in his 2013 film *Her*:** *Her*, written and directed by Spike Jonze (2013; Culver City, CA: Sony Pictures Home Entertainment, 2014).

158 **Sony's newest version of the AIBO:** "Aibo," Sony Electronics, accessed August 30, 2020, https://direct.sony.com/aibo/.

158 **holding them for ransom was a lucrative gig:** Zoë Crew, "A Look at Dog Theft in the Nineteenth-Century Slum," Exploring the Slum, December 18, 2015, https://slumexplorers.wordpress.com/animals/zoe/.

158 **The first commercially prepared dog food:** Dashka Slater, "Who Made That Dog Biscuit?," *New York Times Magazine*, August 1, 2014, https://www.nytimes.com/2014/08/03/magazine/who-made-that-dog-biscuit.html.

158 **American pet spending:** "Pet Industry Market Size & Ownership Statistics," American Pet Products Association.

159   **just a few generations ago, most weddings were nowhere close:** Rebecca Mead, *One Perfect Day: The Selling of the American Wedding* (New York: Penguin, 2008).

159   **"People don't know what they want until you show it to them":** Chunka Mui, "Five Dangerous Lessons to Learn from Steve Jobs," *Forbes*, October 17, 2011, https://www.forbes.com/sites/chunkamui/2011/10/17/five-dangerous-lessons-to-learn-from-steve-jobs/.

160   **the casinos and shopping malls designed by psychologists:** Natasha Dow Schüll, *Addiction by Design: Machine Gambling in Las Vegas* (Princeton, NJ: Princeton University Press, 2014); Douglas Rushkoff, *Coercion: Why We Listen to What "They" Say* (New York: Riverhead Books, 2000).

160   **Weizenbaum changed his mind:** Joseph Weizenbaum, *Computer Power and Human Reason: From Judgment to Calculation* (New York: W. H. Freeman, 1976).

160   **In 2003, legal scholar Ian Kerr foresaw:** Ian R. Kerr, "Bots, Babes and the Californication of Commerce," *University of Ottawa Law & Technology Journal* 1 (2003): 285.

160   **for example, in the form of bots on dating apps:** Leo Kelion, "Tinder Accounts Spammed by Bots Masquerading as Singles," BBC News, April 2, 2014, https://www.bbc.com/news/26850761.

160   **The Campaign for a Commercial-Free Childhood:** Campaign for a Commercial-Free Childhood (website), accessed August 30, 2020, https://commercialfreechildhood.org/.

161   **An early version of the Fitbit had a flower on it:** Jenna Wortham, "Fitbit's Motivator: A Virtual Flower," *New York Times*, December 10, 2009, https://bits.blogs.nytimes.com/2009/12/10/fitbits-motivator-a-virtual-flower/.

161   **Every new media and technology format has the potential:** Rushkoff, *Coercion*.

161   **Woody Hartzog paints a science-fictional scene:** Woodrow Hartzog, "Unfair and Deceptive Robots," *Maryland Law Review* 74, no. 4 (2014): 785.

BIAS IN ARTIFICIAL AGENT DESIGN

162   **Men called in saying they didn't want to take driving directions:** Clifford Nass and Corina Yen, *The Man Who Lied to His Laptop: What We Can Learn about Ourselves from Our Machines* (New York: Penguin, 2010).

163   **even though the computers gave identical information:** Clifford Nass, Youngme Moon, and Nancy Green, "Are Machines Gender Neutral? Gender-Stereotypic Responses to Computers with Voices," *Journal of Applied Social Psychology* 27, no. 10 (1997): 864–76.

163   **consistently perceived as more knowledgeable:** Aaron Powers and Sara Kiesler, "The Advisor Robot: Tracing People's Mental Model from a Robot's Physical Attributes," in *Proceedings of the 1st ACM SIGCHI/SIGART Conference on Human-Robot Interaction* (2006): 218–25.

163   **People will rate a robot with long hair:** Friederike Eyssel and Frank Hegel, "(S)he's Got the Look: Gender Stereotyping of Robots," *Journal of Applied Social Psychology* 42, no. 9 (2012): 2213–30.

163   **Andra Keay surveyed robot names:** Andra Keay, "The Naming of Robots: Biomorphism, Gender and Identity" (master's thesis, University of Sydney, 2012).

163   **Researchers have found that artificial agents with faces:** Clifford Nass and Youngme Moon, "Machines and Mindlessness: Social Responses to Computers," *Journal of Social Issues* 56, no. 1 (2000): 81–103; see also Jean A. Pratt, Karina Hauser, Zsolt Ugray, and Olga Patterson, "Looking at Human–Computer Interface Design: Effects of Ethnicity in Computer Agents," *Interacting with Computers* 19, no. 4 (2007): 512–23.

163   **robots that were given names of the same ethnicity:** Friederike Eyssel and Dieta Kuchenbrandt, "Social Categorization of Social Robots: Anthropomorphism as a Function of Robot Group Membership," *British Journal of Social Psychology* 51, no. 4 (2012): 724–31.

164   **In 2020, the WHO . . . launched an "artificial health worker":** "Using AI to Quit Tobacco," World Health Organization, accessed August 30, 2020, https://www.who.int/news-room/spotlight/using-ai-to-quit-tobacco.

164   **Director Michael Bay brushed off criticism of racial stereotypes:** Sandy Cohen, "Transformers' Jive-Talking Robots Raise Race Issues," *Huffington Post*, July 25, 2009.

164   **ways to create robot characters without gender or racial cues:** Joshua Skewes, David M. Amodio, and Johanna Seibt, "Social Robotics and the Modulation of Social Perception and Bias," *Philosophical Transactions of the Royal Society B* 374, no. 1771 (2019): 20180037.

164   **roboticist Ayanna Howard argues that's what we should do:** Ayanna Howard, *Sex, Race, and Robots: How to be Human in the Age of AI* (Audible Original, 2020).

PRIVACY

165   **"is sympathetic, listens and obviously cannot tell":** Gerald P. Mallon, "Cow as Co-therapist: Utilization of Farm Animals as Therapeutic Aides with Children in Residential Treatment," *Child and Adolescent Social Work Journal* 11, no. 6 (1994): 455–74. (Today, Mallon is the executive director of the National Center for Child Welfare Excellence and a professor at Hunter College.)

165   **so that she could trust ELIZA with her secrets "in private":** Weizenbaum, *Computer Power and Human Reason.*

166   **Hello Barbie:** Samuel Gibbs, "Hackers Can Hijack Wi-Fi Hello Barbie to Spy on Your Children," *Guardian*, November 26, 2015, http://www.theguardian.com/technology/2015/nov/26/hackers-can-hijack-wi-fi-hello-barbie-to-spy-on-your-children; Lauren Walker, "Hello Barbie, Your Child's Chattiest and Riskiest Christmas Present," *Newsweek*, December 15, 2015, https://www

.newsweek.com/2015/12/25/hello-barbie-your-childs-chattiest-and-riskiest
-christmas-present-404897.html; Laura Hautala, "Hello Barbie: She's Just Insecure," *CNET*, December 4, 2015, https://www.cnet.com/news/hello-headaches
-barbie-of-the-internet-age-has-even-more-security-flaws/.

166   **My Friend Cayla:** Soraya Sarhaddi Nelson, "Germany Bans 'My Friend
        Cayla' Doll over Spying Concerns," NPR, February 20, 2017, https://www
        .npr.org/2017/02/20/516292295/germany-bans-my-friend-cayla-doll-over
        -spying-concerns.

167   **Boxie:** Alexander James Reben, "Interactive Physical Agents for Story Gath-
        ering" (PhD diss., Massachusetts Institute of Technology, 2010).

167   **more inclined to give the computer intimate answers to personal ques-
        tions:** Youngme Moon, "Intimate Exchanges: Using Computers to Elicit
        Self-Disclosure from Consumers," *Journal of Consumer Research* 26, no. 4
        (2000): 323–39.

167   **self-disclosure of information increases:** Eric J. Vanman and Arvid Kappas,
        "'Danger, Will Robinson!' The Challenges of Social Robots for Intergroup
        Relations," *Social and Personality Psychology Compass* 13, no. 8 (2019): e12489;
        Gale M. Lucas, Jonathan Gratch, Aisha King, and Louis-Philippe Morency,
        "It's Only a Computer: Virtual Humans Increase Willingness to Disclose,"
        Computers in Human Behavior 37 (2014): 94–100.

167   **can encourage self-disclosure by hugging people:** Masahiro Shiomi, Aya
        Nakata, Masayuki Kanbara, and Norihiro Hagita, "A Robot That Encourages
        Self-Disclosure by Hug," in *International Conference on Social Robotics* (New
        York: Springer, 2017), 324–33.

167   **the collection and aggregation of our personal data:** Bruce Schneier, *Data
        and Goliath: The Hidden Battles to Collect Your Data and Control Your World*
        (New York: W. W. Norton, 2015).

167   **our markets are reducing the human experience to behavioral data that
        companies feed on:** Shoshana Zuboff, *The Age of Surveillance Capitalism:
        The Fight for a Human Future at the New Frontier of Power* (London: Profile
        Books, 2019).

168   **In fact, Fitbit has come under criticism for data collection:** Laura Ryan,
        "Fitbit Hires Lobbyists after Privacy Controversy," *National Journal*, Septem-
        ber 15, 2014.

168   **Stricter privacy settings are often directly at odds:** James Grimmelmann,
        "Saving Facebook," *Iowa Law Review* 94 (2009): 1137.

168   **smart speakers for people's homes rose to 147 million units:** "Global
        Smart Speaker Vendor & OS Shipment and Installed Base Market Share by
        Region: Q4 2019," Strategy Analytics, February 13, 2019.

168   **There was a huge backlash when Facebook did a study:** Blake Hallinan, Jed
        R. Brubaker, and Casey Fiesler, "Unexpected Expectations: Public Reaction
        to the Facebook Emotional Contagion Study," *New Media & Society* 22, no. 6
        (2020): 1076–94.

168   **people's empathy for robots also increases their perceptions of threat:**
        Shelby Ceh and Eric Vanman, "The Robots Are Coming! The Robots Are
        Coming! Fear and Empathy for Human-like Entities," *PsyArXiv* (2018), doi:10
        .31234/osf.io/4cr2u; Jan-Philipp Stein and Peter Ohler, "Venturing into the
        Uncanny Valley of Mind—The Influence of Mind Attribution on the Accep-
        tance of Human-like Characters in a Virtual Reality Setting," *Cognition* 160
        (2017): 43–50.

168   **people will get very fixated on a robot cheating:** Alexandru Litoiu,
        Daniel Ullman, Jason Kim, and Brian Scassellati, "Evidence That Robots
        Trigger a Cheating Detector in Humans," *Proceedings of the 10th ACM/
        IEEE International Conference on Human-Robot Interaction*, Portland, OR
        (2015): 2–5.

THE PUPPET MASTER

169   **look to the puppet master:** Zuboff, *Age of Surveillance Capitalism.*

169   **"celibacy syndrome" and culture in Japan:** Abigail Haworth, "Why Have
        Young People in Japan Stopped Having Sex?," *Guardian*, October 20, 2013,
        http://www.theguardian.com/world/2013/oct/20/young-people-japan
        -stopped-having-sex.

170   **In 2015, a meta-study of seventy different studies:** Julianne Holt-Lunstad,
        Timothy B. Smith, Mark Baker, Tyler Harris, and David Stephenson, "Lone-
        liness and Social Isolation as Risk Factors for Mortality: A Meta-Analytic
        Review," *Perspectives on Psychological Science* 10, no. 2 (2015): 227–37.

170   **harm of loneliness is similar to smoking fifteen cigarettes a day:** Liana
        DesHarnais Bruce, Joshua S. Wu, Stuart L. Lustig, Daniel W. Russell, and Doug-
        las A. Nemecek, "Loneliness in the United States: A 2018 National Panel Sur-
        vey of Demographic, Structural, Cognitive, and Behavioral Characteristics,"
        American Journal of Health Promotion 33, no. 8 (2019): 1123–33; Nick Tate,
        "Loneliness Rivals Obesity, Smoking as Health Risk," WebMD, May 4, 2018,
        https://www.webmd.com/balance/news/20180504/loneliness-rivals-obesity
        -smoking-as-health-risk.

170   **research shows that anthropomorphism can meet our inherent need to
        be social:** Nicholas Epley, Adam Waytz, and John T. Cacioppo, "On Seeing
        Human: A Three-Factor Theory of Anthropomorphism," *Psychological Review*
        114, no. 4 (2007); see also Friederike Eyssel and Natalia Reich, "Loneliness
        Makes the Heart Grow Fonder (of Robots)—on the Effects of Loneliness on
        Psychological Anthropomorphism," in *8th ACM/IEEE International Confer-
        ence on Human-Robot Interaction (HRI)* (2013): 121–22.

170   **"technological solutionism":** Evgeny Morozov, *To Save Everything, Click
        Here: The Folly of Technological Solutionism* (New York: PublicAffairs, 2013).

170   **technologies can inherently embody social and political power:** Winner,
        "Do Artifacts Have Politics?"

170   **Sasha Costanza-Chock . . . makes a powerful case:** Sasha Costanza-Chock, "Design Justice: Towards an Intersectional Feminist Framework for Design Theory and Practice," *Proceedings of the Design Research Society* (2018).

171   **people are more worried about drones capturing visuals:** Ryan Calo, "The Drone as a Privacy Catalyst," *Stanford Law Review* 64 (2011): 29; and in-person conversation.

171   **"animals are good to think with":** Claude Lévi-Strauss, *The Savage Mind* (Chicago: University of Chicago Press, 1966).

III VIOLENCE, EMPATHY, AND RIGHTS

8. WESTERN ANIMAL AND ROBOT RIGHTS THEORIES

175   **Quote:** Douglas Adams, *The Restaurant at the End of the Universe* (London: Tor UK, 1989).

175   **the 2016 TV show *Westworld*:** *Westworld*, created by Jonathan Nolan and Lisa Joy, directed by Richard J. Lewis et al. (HBO, October 2, 2016).

176   **"The [AI] honeymoon is probably going to last for about 18 months":** Larry Fitzmaurice, "How the Creators of 'Westworld' Built a Violent World of Robot Cowboys," *Vice*, September 30, 2016, https://www.vice.com/en_au /article/yvebxw/westworld-jonathan-nolan-lisa-joy-interview.

DO ROBOTS DESERVE RIGHTS?

177   **David Gunkel's book . . . pulls together much of the conversation:** David J. Gunkel, *Robot Rights* (Cambridge, MA: MIT Press, 2018).

177   **what are rights?:** Tom Campbell, *Rights: A Critical Introduction* (Milton Park, UK: Taylor & Francis, 2011); Jeremy Waldron, ed., *Theories of Rights* (Oxford: Oxford University Press, 1985).

178   **"as silly as discriminatory treatment of humans":** Hilary Putnam, "Robots: Machines or Artificially Created Life?," *Journal of Philosophy* 61, no. 21 (1964): 668–91.

178   **from moral tests, to brain cell count, to giving robots rights once they are able to demand them for themselves:** Robert Sparrow, "The Turing Triage Test," *Ethics and Information Technology* 6, no. 4 (2004): 203–13; Kevin Warwick, "Robots with Biological Brains," in *Robot Ethics: The Ethical and Social Implications of Robotics*, ed. Patrick Lin, Keith Abney, and George A. Bekey (Cambridge, MA: MIT Press, 2012), 317–32; Keith Abney, "Robotics, Ethical Theory, and Metaethics: A Guide for the Perplexed," in *Robot Ethics: The Ethical and Social Implications of Robotics*, ed. Patrick Lin, Keith Abney, and George A. Bekey (Cambridge, MA: MIT Press, 2012), 35–52.

178   **err on the side of giving them rights if we're not sure:** Erica L. Neely, "Machines and the Moral Community," *Philosophy & Technology* 27, no. 1 (2014): 97–111.

DO ANIMALS DESERVE RIGHTS?

179   **Pythagoras and vegetarianism:** Marta Zaraska, *Meathooked: The History and Science of Our 2.5-Million-Year Obsession with Meat* (New York: Basic Books, 2016); Porphyry, *Life of Pythagoras*, trans. Kenneth Sylvan Guthrie (Alpine, NJ: Platonist Press, 1920), http://www.tertullian.org/fathers/porphyry_life_of_pythagoras_02_text.htm.

180   **French philosopher René Descartes even argued:** René Descartes, *Selections*, ed. Ralph M. Eaton (New York: Scribner's, 1955).

180   **"the question is not, Can they reason?":** Jeremy Bentham, *An Introduction to the Principles of Morals and Legislation* (1789), ed. J. H. Burns and H. L. A. Hart (London: Athlone Press, 1970).

180   **Darwin's work started making the case:** Charles Darwin, *The Descent of Man and Selection in Relation to Sex*, vol. 1 (New York: D. Appleton, 1896).

181   **"very different in the way that they interact with the world":** Rodney Allen Brooks, "What Is It Like to Be a Robot?," March 18, 2017, https://rodneybrooks.com/what-is-it-like-to-be-a-robot/.

181   **we can't make a computer that feels pain:** Daniel C. Dennett, "Why You Can't Make a Computer That Feels Pain," *Synthese* 38, no. 3 (1978): 415–56.

182   **as we gradually discover more about the underwater animal kingdom:** See, for example, Anthony G. Deakin, Jonathan Buckley, Hamzah S. AlZu'bi, Andrew R. Cossins, Joseph W. Spencer, Waleed Al'Nuaimy, Iain S. Young, Jack S. Thomson, and Lynne U. Sneddon, "Automated Monitoring of Behaviour in Zebrafish after Invasive Procedures," *Scientific Reports* 9, no. 1 (2019): 1–13.

182   **A number of scholars have argued:** See summary in Gunkel, *Robot Rights*; see also Joanna J. Bryson, "Robots Should Be Slaves," in *Close Engagements with Artificial Companions: Key Social, Psychological, Ethical and Design Issues*, ed. Yorick Wilks (Amsterdam: John Benjamins, 2010), 63–74.

182   **the Ameglian Major Cow:** Adams, *The Restaurant at the End of the Universe.*

182   **creating robots that specifically desire to serve us:** Steve Petersen, "Designing People to Serve," in *Robot Ethics: The Ethical and Social Implications of Robotics*, ed. Patrick Lin, Keith Abney, and George A. Bekey (Cambridge, MA: MIT Press, 2012).

THE INDIRECT APPROACH

183   **Immanuel Kant on animal rights:** Immanuel Kant, *Lectures on Ethics*, vol. 2, ed. Peter Heath and J. B. Schneewind, trans. Peter Heath (Cambridge: Cambridge University Press, 2001).

183   **John Locke:** John Locke, *Some Thoughts Concerning Education* (1693; Menston, UK: Scolar Press, 1970).

183   **protect robots from violence for the sake of ourselves:** See, for example, Kate Darling, "Extending Legal Protections to Social Robots: The Effects of Anthropomorphism, Empathy, and Violent Behavior towards Robotic

Objects," in *Robot Law*, ed. Ryan Calo, A. Michael Froomkin, and Ian R. Kerr (Cheltenham, UK: Edward Elgar, 2016); Blay Whitby, "Sometimes It's Hard to Be a Robot: A Call for Action on the Ethics of Abusing Artificial Agents," *Interacting with Computers* 20, no. 3 (2008): 326–33; Anne Gerdes, "The Issue of Moral Consideration in Robot Ethics," *ACM SIGCAS Computers and Society* 45, no. 3 (2016): 274–79; Shannon Vallor, *Technology and the Virtues: A Philosophical Guide to a Future Worth Wanting* (Oxford: Oxford University Press, 2016); Mark Coeckelbergh, "Is It Wrong to Kick a Robot? Towards a Relational and Critical Robot Ethics and Beyond," in *What Social Robots Can and Should Do: Proceedings of Robophilosophy 2016/TRANSOR 2016* 290 (2016): 7; John Danaher, "Welcoming Robots into the Moral Circle: A Defence of Ethical Behaviourism," *Science and Engineering Ethics* 26, no. 4 (2020): 2023–49; Tony J. Prescott, "Robots Are Not Just Tools," *Connection Science* 29, no. 2 (2017): 142–49.

184   **"but I would shoot it cleanly if the opportunity arose":** Nel Noddings, *Caring: A Relational Approach to Ethics and Moral Education* (Berkeley: University of California Press, 2013).

184   **Sven Nyholm criticizes the approach:** Sven Nyholm, *Humans and Robots: Ethics, Agency, and Anthropomorphism* (Lanham, MD: Rowman & Littlefield, 2020).

9. FREE WILLY: WESTERN ANIMAL RIGHTS IN PRACTICE

186   **Quote:** Hal Herzog, *Some We Love, Some We Hate, Some We Eat: Why It's So Hard to Think Straight about Animals* (New York: HarperCollins, 2010).

186   **Quote:** David Attenborough, speech to Communicate conference, November 2010.

187   **lots of healthy omega-3 fatty acids:** Susanna Forrest, "The Troubled History of Horse Meat in America," *Atlantic*, June 8, 2017, https://www.theatlantic .com/technology/archive/2017/06/horse-meat/529665/.

KINDNESS AND CLASSISM

188   **Valluvar's *Thirukkural*:** Gana Nath Das, *Readings from* Thirukkural (Abhinav Publications, 1997).

188   **In the 1600s, the Japanese Tokugawa shogunate enacted animal protection laws:** Hisashi Tsuruoka, "Shoguns and Animals," *Japan Medical Association Journal: JMAJ* 59, no. 1 (2016): 49.

188   **hundreds of years ago, Europeans burned cats alive:** Adrian Franklin, *Animals and Modern Cultures: A Sociology of Human-Animal Relations in Modernity* (London: Sage, 1999).

188   **questioning our exploitative relationship with other creatures:** Keith Thomas, *Man and the Natural World: Changing Attitudes in England, 1500–1800* (London: Penguin UK, 1991).

188   **pets . . . became a catalyst for the upper classes to develop a sense of
       mercy:** Harriet Ritvo, *The Animal Estate: The English and Other Creatures in
       the Victorian Age* (Cambridge, MA: Harvard University Press, 1987).

189   **What would be next, people asked, rights for asses, dogs, and cats?:** Simon
       Brooman and Debbie Legge, *Law Relating to Animals* (London: Cavendish,
       1997).

189   **setting this type of precedent could lead to all sorts of . . . consequences:**
       Franklin, *Animals and Modern Cultures*.

189   **protect animals for the sake of *human* behavior:** Ritvo, *Animal Estate*;
       Thomas G. Kelch, "A Short History of (Mostly) Western Animal Law: Part II,"
       Animal Law 19 (2012): 347; Franklin, *Animals and Modern Cultures*.

189   **good old-fashioned classism:** Franklin, *Animals and Modern Cultures*;
       Ritvo, *Animal Estate*.

189   **Parliaments even voted down some animal protection legislation:** Brooman
       and Legge, *Law Relating to Animals*.

189   **prevent the rich from engaging in their beloved recreational activities:**
       Franklin, *Animals and Modern Cultures*.

190   **the animals in London's Smithfield Market were abused and mistreated:**
       Hilda Kean, *Animal Rights: Political and Social Change in Britain since 1800*
       (London: Reaktion Books, 1998).

190   **"poor harmless milch asses . . . banged by a fellow with a thick cudgel":**
       Report and Proceedings of the Annual Meeting of the Society for the Pre-
       vention of Cruelty to Animals, 1836.

THE ANTI-VIVISECTIONISTS

191   **despite refusing to do experiments on animals:** Samuel Hopgood Hart, "In
       Memoriam Anna Kingsford," Leeds Vegetarian Society, 1947, http://www
       .humanitarismo.com.br/annakingsford/english/Biographical_Information
       _and_Biographies/Texts/IBB-I-HartMemoAKtxt.htm.

191   **Frances Power Cobbe:** Jane Legget, *Local Heroines: A Women's History Gaz-
       etteer of England, Scotland and Wales* (London: Pandora, 1988), 50; David
       Allan Feller, "Dog Fight: Darwin as Animal Advocate in the Antivivisection
       Controversy of 1875," *Studies in History and Philosophy of Science Part C:
       Studies in History and Philosophy of Biological and Biomedical Sciences* 40,
       no. 4 (2009): 265–71.

191   **joined by Queen Victoria:** Kelch, "Western Animal Law."

191   ***A Dog's Tale*:** Mark Twain, *A Dog's Tale* (Harper & Brothers, 1904).

192   **the Brown Dog Affair:** Lizzy Lind af Hageby and Leisa Katherina Schartau,
       The Shambles of Science: Extracts from the Diary of Two Students of Physiology
       (London: Ernest Bell, 1903); Coral Lansbury, *The Old Brown Dog: Women,
       Workers, and Vivisection in Edwardian England* (Madison: University of Wis-
       consin Press, 1985); Peter Mason, *The Brown Dog Affair: The Story of a Monu-
       ment That Divided a Nation* (London: Two Sevens, 1997); "VIVISECTIONIST

EXCULPATED.; Hon. Stephen Coleridge Must Pay Prof. Bayliss 2,000 Damages for Libel," *New York Times*, November 19, 1903, https://www.nytimes.com/1903/11/19/archives/vivisectionist-exculpated-hon-stephen-coleridge-must-pay-prof.html; Lisa Galmark, "Women Antivivisectionists—the Story of Lizzy Lind af Hageby and Leisa Schartau," *Animal Issues* 4, no. 2 (2000): 1; "Brown Dog Statue," Atlas Obscura, accessed August 29, 2020, http://www.atlasobscura.com/places/brown-dog-statue.

193 **leading to the Laboratory Animal Welfare Act of 1966:** Bernard E. Rollin, "The Regulation of Animal Research and the Emergence of Animal Ethics: A Conceptual History," *Theoretical Medicine and Bioethics* 27, no. 4 (2006): 285–304; Thomas G. Kelch, *Globalization and Animal Law: Comparative Law, International Law and International Trade* (Alphen aan den Rijn, NL: Kluwer Law International, 2017).

193 **the American Museum of Natural History in New York City was performing experiments on cats:** Nathaniel Sheppard Jr., "Cats' Mutilation Laid to Museum," *New York Times*, July 19, 1976, https://www.nytimes.com/1976/07/19/archives/cats-mutilation-laid-to-museum-natural-history-is-accused-of.html.

193 *Unnecessary Fuss:* Brian Lowe, "Animal Rights Struggles to Dominate the Public Moral Imagination through Sociological Warfare," *Theory in Action* 1, no. 3 (2008): 1–24.

SELECTIVE EMPATHY AND THE ANGORA PROJECT

194 **lab mice:** Herzog, *Some We Love, Some We Hate, Some We Eat*; Larry Carbone, *What Animals Want: Expertise and Advocacy in Laboratory Animal Welfare Policy* (Oxford: Oxford University Press, 2004).

194 **the idea of biological egalitarianism didn't do much to shift people's thinking:** Ritvo, *Animal Estate*.

194 **Nazi animal rights laws:** Susan Orlean, *Rin Tin Tin: The Life and the Legend*, enhanced ebook (New York: Simon and Schuster, 2011); Kelch, "Western Animal Law," 347.

194 **The Angora project:** "Angora: Pictorial Records of an SS Experiment," *Wisconsin Magazine of History* 50, no. 4 (unpublished documents on Nazi Germany from the Mass Communications History Center, summer 1967): 392–413, https://www.wisconsinhistory.org/Records/Article/CS3952.

THE ANTHROPOMORPHISM CONTROVERSY

195 **movement against the male establishment in medicine and science:** Hilda Kean, "The 'Smooth Cool Men of Science': The Feminist and Socialist Response to Vivisection," *History Workshop Journal* 40, no. 1 (Autumn 1995): 16–38.

196 **other than viewing animals as mechanical was unscientific:** Richard D. Ryder, *Animal Revolution: Changing Attitudes Towards Speciesism* (Oxford: Berg, 2000).

196   "disguising anthropomorphic (in other words, pre-scientific) ways of thinking as science": Roger Scruton, "Animal Rights," *City Journal*, Summer 2000, https://www.city-journal.org/html/animal-rights-11955.html.

196   *Black Beauty*: Anna Sewell, *Black Beauty* (Norwich, UK: Jarrold, 1877).

196   whaling started to become a proper industry: Marissa Fessenden, "We Now Have a Toll of All the Whales Killed by Hunting in the Last Century," *Smithsonian Magazine*, March 12, 2015, https://www.smithsonianmag.com/smart -news/we-now-have-toll-all-whales-killed-hunting-last-century-180954537/.

197   whales . . . saved by the beauty of their song: David Rothenberg, "Nature's Greatest Hit: The Old and New Songs of the Humpback Whale," *Wire*, September 2014, https://www.thewire.co.uk/in writing/essays/nature_s -greatest-hit_the-old-and-new-songs-of-the-humpback-whale; Michael May, "Recordings That Made Waves: The Songs That Saved the Whales," NPR, December 26, 2014, https://www.npr.org/2014/12/26/373303726/recordings -that-made-waves-the-songs-that-saved-the-whales.

197   asked participants to read stories about dogs: See, for example, Max E. Butterfield, Sarah E. Hill, and Charles G. Lord, "Mangy Mutt or Furry Friend? Anthropomorphism Promotes Animal Welfare," *Journal of Experimental Social Psychology* 48, no. 4 (2012): 957–60.

197   people project some of their own personality traits: Christina M. Brown and Julia L. McLean, "Anthropomorphizing Dogs: Projecting One's Own Personality and Consequences for Supporting Animal Rights," *Anthrozoös* 28, no. 1 (2015): 73–86.

198   based on the size of the animal's eyes: Herzog, *Some We Love*.

198   more likely to get adopted from shelters: Bridget M. Waller, Kate Peirce, Cátia C. Caeiro, Linda Scheider, Anne M. Burrows, Sandra McCune, and Juliane Kaminski, "Paedomorphic Facial Expressions Give Dogs a Selective Advantage," *PLOS ONE* 8, no. 12 (2013): e82686.

198   "They stopped the cute baby seal hunt": Herzog, *Some We Love*.

198   anthropomorphism is also controversial within the ranks: Domenica Bruni, Pietro Perconti, and Alessio Plebe, "Anti-anthropomorphism and Its Limits," *Frontiers in Psychology* 9 (2018): 2205; Doris Lin, "Anthropomorphism and Animal Rights" *ThoughtCo*, December 19, 2017, https://www .thoughtco.com/anthropomorphism-and-animal-rights-127579.

199   Over the last few decades, our legal systems have moved closer: Kelch, Western Animal Law," 347.

199   in 2000, a High Court in India ruled: Martha C. Nussbaum, "The Moral Status of Animals," *Chronicle of Higher Education* 52, no. 22 (2006): B6–8.

OUR BIGGEST HYPOCRISY: MEAT

199   Shenzhen banned the sale and consumption of dog and cat meat: "Shenzhen Becomes First Chinese City to Ban Eating Cats and Dogs," BBC News, April 2, 2020, https://www.bbc.com/news/world-asia-china-52131940.

199   **Dog and Cat Meat Trade Prohibition Act:** Dog and Cat Meat Trade Prohibition Act of 2018 (H.R. 6720), also called the DCMTPA.

200   **a dominant role in the world's meat production:** OECD/FAO, "OECD-FAO Agricultural Outlook," OECD Agriculture Statistics, 2019, http://dx.doi.org/10.1787/agr-outl-data-en.

200   **Meat consumption has doubled:** OECD/FAO, "Meat and Meat Products," Animal Production and Health Division, March 15, 2019, http://www.fao.org/ag/againfo/themes/en/meat/home.html.

200   **"the single most severe, systematic example of man's inhumanity to another sentient animal":** Peter Singer and Jim Mason, "Are We What We Eat?," *Soundings* 34 (2006): 67–72.

200   **McDonald's pledging to switch to cage-free eggs by 2025:** Beth Kowitt, "Inside McDonald's Bold Decision to Go Cage-Free," *Fortune,* accessed August 31, 2020, https://fortune.com/longform/mcdonalds-cage-free/.

200   **Isabella Beeton advocated against caging hens in 1861:** Isabella Beeton and Teresa Andrews, *Considering Chickens* (n.p.: Forest House, 2019).

201   **eat meat or wear leather without batting an eye:** Richard W. Bulliet, *Hunters, Herders, and Hamburgers: The Past and Future of Human-Animal Relationships* (New York: Columbia University Press, 2005).

201   **cockfighting and horse racing:** John Cherwa, "Santa Anita Suffers 14th Horse Death since Start of Season," *Los Angeles Times,* June 7, 2020, https://www.latimes.com/sports/story/2020-06-07/santa-anita-suffers-14-horse-racing-deaths-start-season; Herzog, *Some We Love.*

201   **"We know that animals' behavior and physical attributes play a powerful role":** Brian Fagan, *The Intimate Bond: How Animals Shaped Human History* (New York: Bloomsbury, 2015).

201   **we judge each other for our choices:** For example, Japan has opposed restrictions on whale hunting; see Bulliet, *Hunters, Herders, and Hamburgers.*

201   **"vast, complicated, and, for the most part, boring":** Herzog, *Some We Love.*

10. DON'T KICK THE ROBOT

203   **Quote:** Frans B. M. de Waal, "Are We in Anthropodenial?," *Discover,* January 18, 1997, 50–53.

EMPATHY FOR ROBOTS

205   **Freedom Baird and the podcast experiment:** WNYC Studios, "Furbidden Knowledge," *Radiolab,* May 31, 2011, https://www.wnycstudios.org/podcasts/radiolab/segments/137469-furbidden-knowledge.

207   **a key component of our social interactions:** Stephanie D. Preston and Frans B. M. de Waal, "Empathy: Its Ultimate and Proximate Bases," *Behavioral and Brain Sciences* 25, no. 1 (2002): 1–20.

211   **"Is It Cruel to Kick a Robot Dog?":** Phoebe Parke, "Is It Cruel to Kick a Robot Dog?," CNN, February 13, 2015, https://www.cnn.com/2015/02/13/tech/spot-robot-dog-google/index.html.

TRAINING OUR CRUELTY MUSCLES?

212   **YouTube also took down a bunch of robot-on-robot violence videos:** Victor Tangermann, "YouTube's AI Has Been Flagging Robot Battle Videos For 'Animal Cruelty,'" *ScienceAlert*, August 22, 2019, https://www.sciencealert.com/youtube-s-ai-algorithm-has-been-removing-robot-battle-videos-for-animal-cruelty.

212   **blog post published by *WIRED* in 2009:** Daniel Roth, "Do Humanlike Machines Deserve Human Rights?," *WIRED*, January 19, 2009, https://www.wired.com/2009/01/st-essay-16/.

SEX, VIDEO GAMES, AND ANIMALS

214   **Humanoid sex robot hype and reality:** Chelsea G. Summers, "There Are a Lot of Problems with Sex Robots," *OneZero*, Medium, July 26, 2018, https://onezero.medium.com/there-are-a-lot-of-problems-with-sex-robots-38ea0c17b7db; Kate Devlin, *Turned On: Science, Sex and Robots* (New York: Bloomsbury, 2018).

215   **worried that men will begin to prefer sex with sex robots:** Kathleen Richardson, "Sex Robot Matters: Slavery, the Prostituted, and the Rights of Machines," *IEEE Technology and Society Magazine* 35, no. 2 (2016): 46–53.

215   **desensitize people to . . . behaviors that we don't want to encourage:** see, for example, Jeannie Suk Gersen, "Sex Lex Machina: Intimacy and Artificial Intelligence," *Columbia Law Review* 119, no. 7 (2019): 1793–1810; Richardson, "Sex Robot Matters," 46–53.

215   **has ruled that virtual child pornography is legal:** Ashcroft v. Free Speech Coalition (00-795), 535 US 234 (2002) 198 F.3d 1083, affirmed.

215   **the show *Westworld* is heavily influenced by modern video games:** Larry Fitzmaurice, "How the Creators of 'Westworld' Built a Violent World of Robot Cowboys," *Vice*, September 30, 2016, https://www.vice.com/en_au/article/yvebxw/westworld-jonathan-nolan-lisa-joy-interview.

216   **Some research finds connections:** Craig A. Anderson and Brad J. Bushman, "Effects of Violent Video Games on Aggressive Behavior, Aggressive Cognition, Aggressive Affect, Physiological Arousal, and Prosocial Behavior: A Meta-Analytic Review of the Scientific Literature," *Psychological Science* 12, no. 5 (2001): 353–59; John L. Sherry, "The Effects of Violent Video Games on Aggression: A Meta-Analysis," *Human Communication Research* 27, no. 3 (2001): 409–31; Anna T. Prescott, James D. Sargent, and Jay G. Hull, "Meta-analysis of the Relationship between Violent Video Game Play and Physical Aggression over Time," *Proceedings of the National Academy of Sciences* 115, no. 40 (2018): 9882–888.

216   **other research finds no relationship:** Christopher J. Ferguson and John Kil-
       burn, "The Public Health Risks of Media Violence: A Meta-Analytic Review,"
       Journal of Pediatrics 154, no. 5 (2009): 759–63; Christopher J. Ferguson, "Do
       Angry Birds Make for Angry Children? A Meta-Analysis of Video Game Influ-
       ences on Children's and Adolescents' Aggression, Mental Health, Prosocial
       Behavior, and Academic Performance," *Perspectives on Psychological Science*
       10, no. 5 (2015): 646–66; Aaron Drummond, James D. Sauer, and Christo-
       pher J. Ferguson, "Do Longitudinal Studies Support Long-Term Relationships
       between Aggressive Game Play and Youth Aggressive Behaviour? A Meta-
       Analytic Examination," *Royal Society Open Science* 7, no. 7 (2020): 200373.

216   **creating some controversy among scientists:** Janie Graziani, "Letter to
       APA on Policy Statement on Violent Media," *Stetson Today*, October 7, 2013,
       https://www.stetson.edu/today/2013/10/letter-to-apa-on-policy-statement
       -on-violent-media/.

216   **their latest statement acknowledges some small:** "APA Reaffirms Position
       on Violent Video Games and Violent Behavior," American Psychological
       Association, https://www.apa.org/news/press/releases/2020/03/violent-video
       -games-behavior.

216   **Politicians and parents have turned video games into an easy scapegoat:**
       Patrick M. Markey and Christopher J. Ferguson, *Moral Combat: Why the
       War on Violent Video Games Is Wrong* (Dallas: BenBella Books, 2017); Karen
       Sternheimer, *It's Not the Media: The Truth about Pop Culture's Influence on
       Children* (New York: Basic Books, 2003).

216   **The National Rifle Association . . . has even blamed video games:** Daniel
       Beekman, "NRA Blames Video Games Like 'Kindergarten Killer' for Sandy
       Hook Elementary School Slaughter," *New York Daily News*, December 21,
       2012, https://www.nydailynews.com/news/national/nra-blames-video-games
       -kindergarten-killer-sandy-hook-article-1.1225212.

216   **Fredric Wertham warned against the dangers of comic books' influence:**
       Fredric Wertham, *Seduction of the Innocent* (New York: Rinehart, 1954).

216   **to affect a US congressional inquiry:** Amy Kiste Nyberg, "Comics
       Code History: The Seal of Approval," Comic Book Legal Defense Fund,
       accessed August 31, 2020, http://cbldf.org/comics-code-history-the-seal
       -of-approval/.

216   **In 2019, China prohibited video games that depict:** Paolo Zialcita, "China
       Introduces Restrictions on Video Games for Minors," NPR, November 6, 2019,
       https://www.npr.org/2019/11/06/776840260/china-introduces-restrictions
       -on-video-games-for-minors.

216   **Some countries . . . restrict the sale of violent video games to minors:**
       "Europe to Ban Sales of Violent Video Games to Kids," Pinsent Masons, January
       18, 2007, https://www.pinsentmasons.com/out-law/news/europe-to-ban-sales
       -of-violent-video-games-to-kids.

216   **the US Supreme Court ended up ruling the bans as unconstitutional:**
       Brown v. Entertainment Merchants Association, 564 U.S. 786 (2011).

217   **cruelty toward animals and cruelty toward children were so closely related:** Thomas G. Kelch, "A Short History of (Mostly) Western Animal Law: Part II," *Animal Law* 19 (2012): 347.

217   **there are cross-reporting laws that require social workers, vets, and doctors:** See, for example, Code of the District of Columbia, § 4-1321.02, "Persons required to make reports procedure"; 510 Illinois Compiled Statutes 70/18, Sec. 18, "Cross-reporting"; Louisiana State Legislature, RS 14 §403.6, "Reporting of neglect or abuse of animals." Other states with cross-reporting of animal and child abuse include California, Colorado, Connecticut, Maine, Massachusetts, Nebraska, Ohio, Tennessee, Virginia, and West Virginia; see "Cross-Reporting of Animal and Child Abuse," American Veterinary Medical Association, April 2018, https://www.avma.org/advocacy/state-local -issues/cross-reporting-animal-and-child-abuse.

217   **Violence toward animals has also been connected to domestic abuse:** Linda Jean Nebbe, "The Human-Animal Bond's Role with the Abused Child," (PhD diss., Iowa State University, 1997); Randall Lockwood and Frank R. Ascione, eds., *Cruelty to Animals and Interpersonal Violence: Readings in Research and Application*, vol. 14 (West Lafayette, IN: Purdue University Press, 1997); Frank R. Ascione and Phil Arkow, eds., *Child Abuse, Domestic Violence, and Animal Abuse: Linking the Circles of Compassion for Prevention and Intervention* (West Lafayette, IN: Purdue University Press, 1999).

217   **A new field called veterinary forensics:** Stefanie Marsh, "The Link Between Animal Abuse and Murder," *Atlantic*, August 31, 2017, https://www.theatlantic .com/science/archive/2017/08/melinda-merck-veterinary-forensics/538575/; see also Randall Lockwood and Phil Arkow, "Animal Abuse and Interpersonal Violence: The Cruelty Connection and Its Implications for Veterinary Pathology," *Veterinary Pathology* 53, no. 5 (2016): 910–18.

217   **Pet and Women Safety (PAWS) Act in the United States:** H.R.909-Pet and Women Safety Act of 2017.

218   **Hexbug empathy study:** Kate Darling, Palash Nandy, and Cynthia Breazeal, "Empathic Concern and the Effect of Stories in Human-Robot Interaction," in *24th IEEE International Symposium on Robot and Human Interactive Communication (RO-MAN)* (2015): 770–75.

219   **which robots they'd be most likely to save in an earthquake:** Laurel D. Riek, Tal-Chen Rabinowitch, Bhismadev Chakrabarti, and Peter Robinson, "How Anthropomorphism Affects Empathy toward Robots," in *Proceedings of the 4th ACM/IEEE International Conference on Human Robot Interaction* (2009): 245–46.

219   **made people watch robot "torture" videos:** Astrid M. Rosenthal-von der Pütten, Nicole C. Krämer, Laura Hoffmann, Sabrina Sobieraj, and Sabrina C. Eimler, "An Experimental Study on Emotional Reactions towards a Robot," *International Journal of Social Robotics* 5, no. 1 (2013): 17–34.

219   **robot and human hands having their fingers cut off:** Yutaka Suzuki, Lisa Galli, Ayaka Ikeda, Shoji Itakura, and Michiteru Kitazaki, "Measuring

Empathy for Human and Robot Hand Pain Using Electroencephalography,"
Scientific Reports 5 (2015): 15924.

219   **empathy for a robot vacuum cleaner that was being verbally harassed:** Mat-
thias Hoenen, Katrin T. Lübke, and Bettina M. Pause, "Non-anthropomorphic
Robots as Social Entities on a Neurophysiological Level," *Computers in
Human Behavior* 57 (2016): 182–86.

A REFLECTION OF OUR HUMANITY

220   **As animal trainer Vicki Hearne has said:** Donna Haraway, *The Companion
Species Manifesto: Dogs, People, and Significant Otherness*, vol. 1 (Chicago:
Prickly Paradigm Press, 2003).

220   **Judith Donath, in her essay "The Robot Dog Fetches for Whom?":** Judith
Donath, "The Robot Dog Fetches for Whom?," in *A Networked Self and
Human Augmentics, Artificial Intelligence, Sentience*, ed. Zizi Papacharissi
(London: Routledge, 2019), 10–24.

THE EMPATHY CONTROVERSY

222   **liking killer whales once they're rebranded to orca:** Jason M. Colby, *Orca:
How We Came to Know and Love the Ocean's Greatest Predator* (Oxford:
Oxford University Press, 2018).

222   **As Paul Bloom has argued:** Paul Bloom, *Against Empathy: The Case for
Rational Compassion* (New York: Random House, 2017).

223   **Philosopher Mark Coeckelbergh also points out:** Mark Coeckelbergh,
"Moral Appearances: Emotions, Robots, and Human Morality," *Ethics and
Information Technology* 12, no. 3 (2010): 235–41.

224   **The anthropomorphism in animal science controversy:** John S. Kennedy,
The New Anthropomorphism (Cambridge: Cambridge University Press, 1992);
Alexandra C. Horowitz and Marc Bekoff, "Naturalizing Anthropomorphism:
Behavioral Prompts to Our Humanizing of Animals," *Anthrozoös* 20, no. 1
(2007): 23–35; Domenica Bruni, Pietro Perconti, and Alessio Plebe, "Anti-
anthropomorphism and Its Limits," *Frontiers in Psychology* 9 (2018): 2205.

224   **animal rights proponent Martha Nussbaum argues:** Martha C. Nussbaum,
Upheavals of Thought: The Intelligence of Emotions (Cambridge: Cambridge
University Press, 2003).

224   **any empathy toward them is wasted narcissism:** See, for example, Joanna
J. Bryson, "Robots Should Be Slaves," in *Close Engagements with Artificial
Companions: Key Social, Psychological, Ethical and Design Issues*, ed. Yorick
Wilks (Amsterdam: John Benjamins, 2010), 63–74.

225   **people's appetite for doling out charity can be easily overwhelmed:** See,
for example, Daniel Västfjäll, Paul Slovic, Marcus Mayorga, and Ellen Peters,
"Compassion Fade: Affect and Charity Are Greatest for a Single Child in
Need," *PLOS ONE* 9, no. 6 (2014): e100115.

225   **A 2019 study by Yon Soo Park and Benjamin Valentino:** Yon Soo Park and
        Benjamin Valentino, "Who Supports Animal Rights? Here's What We Found,"
        Washington Post, July 26, 2019, https://www.washingtonpost.com/politics/2019
        /07/26/who-supports-animal-rights-heres-what-we-found/; Yon Soo Park and
        Benjamin Valentino, "Animals Are People Too: Explaining Variation in Respect
        for Animal Rights," *Human Rights Quarterly* 41, no. 1 (2019): 39–65.

225   **our relationships with companion animals would be meaningless:** James
        A. Serpell, "Anthropomorphism and Anthropomorphic Selection—Beyond
        the 'Cute Response,'" *Society & Animals* 11, no. 1 (2003): 83–100.

FINAL THOUGHTS: PREDICTING THE FUTURE

226   **"just way dumber than they think they are":** Maureen Dowd, "Elon Musk,
        Blasting Off in Domestic Bliss," *New York Times*, July 25, 2020, https://www
        .nytimes.com/2020/07/25/style/elon-musk-maureen-dowd.html.

226   **"a lot of things that we are terribly mistaken about":** Maciej Cegłowski,
        "Superintelligence: The Idea That Eats Smart People," Keynote: WebCamp Zagreb
        Conference 2016, http://2016.webcampzg.org/talks/view/superintelligence-the
        -idea-that-eats-smart-people/.

226   **that he gave it to a graduate student to do over the summer:** William J.
        Broad, "Computer Scientists Stymied in Their Quest to Match Human
        Vision," *New York Times,* September 25, 1984, https://www.nytimes.com
        /1984/09/25/science/computer-scientists-stymied-in-their-quest-to-match
        -human-vision.html.

226   **Toyota announced that by 2010, we would have humanoid robots that
        could:** Ray Kurzweil, "Toyota Aims to Sell Service Robots by 2010," May 31,
        2005, https://www.kurzweilai.net/toyota-aims-to-sell-service-robots-by-2010.

228   **Global sales of industrial robots are projected to grow by 12 percent per
        year:** "Executive Summary World Robotics 2019 Industrial Robots," Inter-
        national Federation of Robotics, accessed August 31, 2020, https://ifr.org
        /downloads/press2018/Executive%20Summary%20WR%202019%20Indus
        trial%20Robots.pdf.

228   **the International Federation of Robotics expects companies to sell over
        68 million robots:** International Federation of Robotics (website), accessed
        August 31, 2020, https://ifr.org/downloads/press2018/Executive_Summary
        _WR_Service_Robots_2019.pdf.

228   **play the "Happy Birthday" tune to itself on Mars:** "Mars Rover Sings
        'Happy Birthday' to Itself," ABC13 Houston, August 5, 2017, https://abc13
        .com/2278350/.

229   **Many states in the US now:** Jeffrey D. Green, Maureen A. Mathews, and
        Craig A. Foster, "Another Kind of 'Interpersonal' Relationship: Humans,
        Companion Animals, and Attachment Theory," *Relationships and Psychol-
        ogy: A Practical Guide* (2009): 87–108.

ACKNOWLEDGMENTS

First and foremost, this book would not have been possible without the support of my husband, Gregory Leppert, who stepped up to be the primary caregiver for our child for five months during the COVID-19 pandemic while I wrote this book. I am also eternally grateful to my son, Benefox Darling, who interrupted my writing process with smiles, hugs, and gifts of rocks that he had collected outdoors. Many thanks to my editor, Serena Jones, for her unwavering support and for believing in me. Chia Evers and Anna Nowogrodzki provided invaluable research assistance during the writing process, and I owe my friend and science journalist Chris Couch the world for editing and fact-checking a large chunk of the manuscript. (Any errors are mine, not theirs.) This book benefited immensely from the thoughts, comments, and corrections of early draft readers Jamie Boyle, Bruce Schneier, Jeanne Darling, Jonathan Zittrain, and Sara Watson. Thank you so much to the hundreds of friends, colleagues, strangers, researchers, designers, company founders, executives, government representatives, robots, and animals I got to meet and learn from in the course of my research. Finally, I would like to thank Teddie Natural Peanut Butter for comprising 80 percent of my food intake while writing.

INDEX

Note: Page numbers in italics indicate figures.

ABOUT THE AUTHOR

DR. KATE DARLING is a leading expert in robot ethics and policy. She's a researcher at the Massachusetts Institute of Technology Media Lab as well as a former fellow at the Harvard Berkman Klein Center for Internet & Society and the Yale Information Society Project. Darling's work has been featured in the *New Yorker,* the *Guardian,* the *Boston Globe, WIRED, Slate,* on PBS and NPR, and more.